Epigenetic Mechanisms of the Cambrian Explosion

Epigenetic Mechanisms of the Cambrian Explosion

Nelson R. Cabej

ACADEMIC PRESS

An imprint of Elsevier

Academic Press is an imprint of Elsevier
125 London Wall, London EC2Y 5AS, United Kingdom
525 B Street, Suite 1650, San Diego, CA 92101, United States
50 Hampshire Street, 5th Floor, Cambridge, MA 02139, United States
The Boulevard, Langford Lane, Kidlington, Oxford OX5 1GB, United Kingdom

Notices
Knowledge and best practice in this field are constantly changing. As new research and experience broaden our understanding, changes in research methods, professional practices, or medical treatment may become necessary.

Practitioners and researchers must always rely on their own experience and knowledge in evaluating and using any information, methods, compounds, or experiments described herein. In using such information or methods they should be mindful of their own safety and the safety of others, including parties for whom they have a professional responsibility.

To the fullest extent of the law, neither the Publisher nor the authors, contributors, or editors, assume any liability for any injury and/or damage to persons or property as a matter of products liability, negligence or otherwise, or from any use or operation of any methods, products, instructions, or ideas contained in the material herein.

Library of Congress Cataloging-in-Publication Data
A catalog record for this book is available from the Library of Congress

British Library Cataloguing-in-Publication Data
A catalogue record for this book is available from the British Library

ISBN: 978-0-12-814311-7

For information on all Academic Press publications visit our
website at https://www.elsevier.com/books-and-journals

Publisher: Charlotte Cockle
Acquisition Editor: Anna Valutkevich
Editorial Project Manager: Kelsey Connors
Production Project Manager: Punithavathy Govindaradjane
Cover Designer: Christian Bilbow

Typeset by TNQ Technologies

Contents

3. Epigenetic requisites of the Cambrian explosion

5. Epilogue

Introduction

The Cambrian explosion is a unique and puzzling event in the history of life on earth, characterized by an eruptive radiation of metazoan forms, unlike, and unrivaled by, any other period in the history of life. It occurred after a long stasis in the evolution of metazoans, within a geologically and evolutionarily very short period of time.

Darwin saw the abruptness of appearance of the breathtaking diversification of forms during the Cambrian as a challenge to his theory. In *The Origin*'s subsection *On the sudden appearance of groups of Allied Species in the lowest known fossiliferous strata*, he emphasized: "There is another and allied difficulty, which is much graver. I allude to the manner in which numbers of species of the same group, suddenly appear in the lowest known fossiliferous rocks." (Darwin, 1859a). However, unwavering in his conviction, Darwin hypothesized that the apparent suddenness and diversity of forms during the Cambrian may be an illusion, resulting from the incomplete fossil record of his time: "I look at the natural geological record, as a history of the world imperfectly kept, and written in a changing dialect; of this history we possess the last volume alone, relating only to two or three countries. Of this volume, only here and there a short chapter has been preserved; and of each page, only here and there a few lines" (Darwin, 1859b).

Nevertheless, now, 150 years after *The Origin*, when an incomparably larger stock of animal fossils has been collected, Darwin's gap remains, the abrupt appearance of Cambrian fossils is a reality, and we are still wondering about the forces and mechanisms that drove it.

Despite the fact that, from time to time, a small number of students have questioned the reality of the Cambrian explosion on the same ground as Darwin, today's consensus is that Cambrian explosion is a scientific fact (Linnemann et al., 2019) and "The Cambrian explosion is real and its consequences set in motion a sea-change in evolutionary history" (Conway Morris, 2000; Nichols et al., 2006).

It marks the beginning of macroevolution (Butterfield, 2007), evolution at the level of taxa, and even Bauplaene rather than molecular level. It started ~542 Ma (million years ago) extending for a period of ~30 million years. But, within ~20 million years (Marshall, 2006; Budd, 2013), it produced that incredible diversity of animal forms, including almost all the extant and several extinct phyla: "Part of the intrigue with the Cambrian explosion is that numerous animal phyla with very distinct body plans arrive on the scene in a geological blink of an eye, with little or no warning of what is to come in rocks that predate this interval of time" (Peterson et al., 2009).

The eruptive character of the Cambrian diversification obviously excludes gene mutations, gene recombination, and drift as its possible causes. Basically, it was rapid evolution of bilaterians, while other great metazoan clades, sponges, placozoans, and cnidarians sat on the bench. The interpretation of this scientifically proven event is of fundamental importance for understanding the nature and the real drivers of eumetazoan evolution.

Despite the accumulation of an immense fossil record, the development of a relevant theoretical groundwork, and the numerous attempts to deal with the causal basis of the Cambrian explosion, just like in Darwin's time, it continues to be one of the greatest enigmas of modern biology.

Now, 110 years after the discovery of the Burger Shale Lagerstätte (Canada), the Cambrian fossil fauna is complemented with similar fossils from other regions around the world (Greenland, Russia, South Africa, United States, etc.), especially the fossil-rich Chengjiang Lägerstatte in Yunnan Province China.

The early Cambrian is characterized by trace fossils and mineralized skeletons of bilaterian animals whose relationship with extant taxa often has been difficult to determine. The conventional marker of the Ediacaran—Cambrian border are considered complex burrowing trace fossils of ichnospecies *Treptichnus pedum* Seilacher, 1955 (Erwin and Valentine, 2013). Fossils of skeletal animals appear in three pulses during the lower Cambrian (c.541—509) (Maloof et al., 2010).

Difficulties often arise in determining the fossil age. Due to variation in time and among lineages, the molecular clock estimates on divergence times are less reliable. They lead to extreme variabilities caused by many sources of error and uncertainties resulting from extrapolations used in determining node ages (dos Reis et al., 2015). So, e.g., the estimates for the age of last eukaryote common ancestor in various studies vary to a hardly acceptable degree, from 1007 (943—1102) Ma to 1898 (1655—2094) Ma (Eme et al., 2014). Even deniers of the Cambrian explosion are compelled to admit: "Molecular dates can be misleading ….The molecular clock is unlikely ever to replace the fossil record as the primary source of information on evolution in deep time. But it has a critical role to play as an alternative historical narrative, potentially complementing the biases and gaps of the paleontological record" (Lindell Bromham, 2009). In view of the conflicting results between the estimates of the time of divergence between taxa obtained by molecular clocks and the fossil data, for the sake of simplicity and clarity, the chronology of evolutionary events in this work will be commonly based on fossil estimates.

Hypotheses on the causes of the Cambrian explosion

Many hypotheses have been presented on the causal basis of the Cambrian explosion. Today, we are still grappling with the question, but no closer to understanding the nature and causes of the Cambrian explosion (Conway Morris, 2000). The various ecological (extinction of the Ediacaran biota, advent of macropredation, bioturbation, etc.) and environmental (rise of temperature at the end of glacial periods, increase of the contents of oxygen in Earth atmosphere, changes in sea salinity and carbon—phosphorus ratio, etc.) hypotheses as causal to the Cambrian Evolution are reviewed briefly.

Most of the alleged causes can certainly play a role in the positive selection of heritable phenotypic changes, but none addresses the central question of the mechanism of induction of the heritable change, which is the essence kernel of the evolutionary process. None of the mechanisms suggested as causative agents of the Cambrian radiation have any visible role as inducer of evolutionary change; none of them addresses the question "How the supposed cause could *generate* the heritable changes?" None of them explains how the change in the environmental temperature, oxygen content, sea salt, even predation, etc., might lead to evolutionary changes and formation of new species or Bauplan.

All that proposed mechanisms can do is create conditions for the action of natural selection or accelerate its action. Natural selection, as an agent of evolution, is relevant as far as the *fate of the change* is concerned, but it is not involved in the generation of the change. In the beginning is change, selection automatically follows it. The change not only precedes but also enables its own selection. From an evolutionary viewpoint, change is the *raison d'être* of selection.

We need to know what is behind the "intriguing mismatches between genomic architecture and body plan complexity" (Conway Morris, 2000). Most of the genes that are related to the emergence of new Bauplan and novel structures existed before the evolution of structures. Two decades ago, Conway Morris emphasized the need to focus not so much on the role of the genetic toolbox but, rather, on the question "how such toolboxes are recruited" (Conway Morris, 2000). We need to address the key question: "Which is the 'user of the genetic toolkit'?" determining expression and recruitment of genes and gene regulatory networks (GRNs) during phylogeny and how basically the same genetic toolkit generated the breathtaking diversity of forms during the Cambrian.

For all the above reasons, I will focus on the inherent evolutionary potentialities of metazoans, especially in the epigenetic mechanisms rather than on external physical and ecological agents involved in evolution of the Cambrian explosion. The existence of an epigenetic system of inheritance was predicted by J. Maynard Smith (Maynard Smith, 1990).

Expression of genes is a central question in biology, including cell differentiation, development, and evolution of animals. There is abundant evidence that gene expression is a function of epigenetic processes, such as DNA methylation/demethylation, histone modification/chromatin remodeling, gene splicing, expression patterns of miRNAs, etc. The fact that all these processes do not occur randomly or casuistically, but instead in strictly determined sites and times, indicates that information of some kind, but obviously different from the genetic information for protein biosynthesis, is used to produce them. Our search for the information determining the occurrence of these epigenetic processes "in the right place at the right time" logically begins with a survey on their proximate causes. Such a survey shows that signals for activating epigenetic processes, in a scientifically adequate number of cases, are of neural origin.

Earlier it was suggested that the evolution of the neuron might have triggered the Cambrian explosion (Stanley, 1992) and recently G.E. Budd expressed the idea that the Cambrian radiation is related to the evolution of the nervous system: "The origins and diversification of the animals, a series of events that became manifest in the so-called 'Cambrian explosion' of ca 540 Ma, must necessarily be intimately tied into the evolution of their important organ systems. Of these, *the nervous system must be considered to be of extreme importance* (emphasis - mine), not only because of its universality among animals apart from sponges and placozoans, but also because of the role it plays in coordination, sensing and indeed many other aspects of the life of an animal" (Budd, 2015). I have argued and adequately substantiated my view on the role of the epigenetic factors and the nervous system in the evolution of metazoans in my earlier work (Cabej, 2012).

The rationale behind the explanation chosen in this work

The failure of the attempts to explain the Cambrian explosion with external (ecological and environmental) factors left us with no choice but to look for causes that are intrinsic to the metazoan themselves as its possible triggers and drivers.

The erection of the metazoan structure, as an improbable structure, requires an immense volume of information, compared to which the information contained in even the most complex genomes represents but a negligible fraction. So, e.g., the human brain alone contains about one trillion neurons (Kandel, 2000). The amount of information necessary only to establish the specific (*nonrandom*) neuronal connections between neurons in the human brain amounts to ~ 1 quadrillion bit (each neuron establishes an average of 10,000 synaptic connections). This exceeds millions of times the information of a few billion bits contained in the nucleotide sequences of DNA in the human genome. Besides, the genetic information comprised in the sequence of DNA base pairs is responsible for the primary structure of proteins, which is different from the information used to determine the highly specific spatial arrangement of billions/trillions of cells of tens to different cell types in the metazoan structure. Being quantitatively negligible and qualitatively ill-suited, the genetic information cannot be taken into account as a possible determinant the metazoan structure. It seems to me that this approach may account for the induction of GRNs in the hypothesis of developmental GRNs.

In this work, the Cambrian explosion will be considered from a causal viewpoint related to the epigenetic theory of evolution. The concept of epigenetics in this work is used in a nonconventional broader meaning to encompass not only epigenetic marks in DNA and RNA but also any other processes involved in the metazoan inheritance. These include the ordered placement of parental cytoplasmic factors in the egg and sperm that control and regulate the expression of zygotic genes and early embryonic development, neural mechanisms of gene splicing and patterns of miRNA expression, gene imprinting, neural mechanisms of activation of tissue-specific/non-housekeeping genes and GRNs, cell differentiation, organogenesis, and morphogenesis in the process of individual development, as well as the neural mechanisms of homeostasis. Within the scope of the work are also mechanisms of the control of the developmental plasticity, intra- and transgenerational inheritance, and evolutionary change (including the speciation process) involving no changes in genes or genetic information in general.

Certainly, the genetic toolkit is involved essentially in the determination of the Bauplan. While necessary, it is not sufficient, for the Bauplan to arise, and we need to know whether its genes are possessors or conveyors of the information or instructions used in erecting the metazoan structure. A closer look at the genetic toolkit shows that its genes are induced by external signals. The fact that external signals occur "in the right place at the right time" rather than randomly indicates that they are consumers of information entering the cell. The conveyor of the information may be a hormone, a growth factor, secreted protein, neuromodulator, neurotransmitter, etc. Typically, a hormonal ligand, as a *primary messenger* of information, binds a specific cell membrane or nuclear receptor. In the typical case, the binding of a hormone to its membrane receptor produces changes in the tertiary and quaternary structure of the receptor, which in this form induces a second messenger (AMP, cyclic GMP, inositol, etc.), which via a chain of phosphorylating reactions amplifies the externally provided information, ultimately leading to the expression of a specific gene. But transmission of the extracellular information to the cell nucleus and the genome is but a link in the chain of information transfer. Tracing back the causal chain, we observe that hormones and endocrine glands secreting them do not "know" the "right time" to secrete the hormone. Indeed, the gland is induced to secrete hormones by specific signals: hormones released by the pituitary. Further upstream, the pituitary in turn is induced by specific hypothalamic signals, neurohormones, and other neural signals to synthesize its hormones. The hypothalamus itself receives input of chemical and electrical

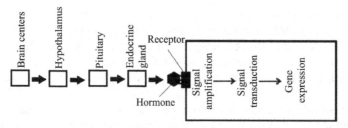

FIGURE 1 Simplified diagrammatic representation of flow of information for gene expression in a generalized hormonal signal cascade. The temporal order of the events in the supracellular signal cascade and the cell signal transduction pathway represents the causal chain from the brain centers to the gene.

signals from other areas of the brain. By processing that input, the hypothalamus produces its output, which is a neurohormone (Fig. 1).

Thus, by tracing back the flow of information from the expressed gene through the transduction pathway to the hormone receptor and all the way back upstream, we reach the brain as the ultimate source of information for the expression of the nonhousekeeping gene. The answer for the anticipated question about the nerveless placozoans and sponges is addressed later in the sections devoted to these clades.

A principle of uniformitarianism and conservation of signaling pathways would predict that the mechanisms of gene expression of extant metazoan clades have been operational during the Cambrian explosion, and mechanisms of the development of extant animals have evolved at same point of their evolutionary history. Notwithstanding the time of origin, these mechanisms also embody elements of the earlier stages of evolution because the evolutionary history of species is built on earlier stages of its evolution. Hence, the study of the mechanisms of the development of extant animals may shed light on the earlier stages of the evolution of extant species. While denying Haeckel's principle that ontogeny is a recapitulation of phylogeny, we are compelled to use his concepts underlying the ontogenetic arguments in determining morphological homologies in metazoans (Alberch and Blanco, 1996) and we are convinced that evolutionary changes are produced during ontogeny: "Ontogenies evolve, not genes. Mutated genes are passed on only to the extent that they promote survival of ontogenies; adulthood is only a fragment of ontogeny" (McKinney and Gittlemann, 1995).

The species ontogeny is a chronicle, albeit blurred and fragmentary, of its evolutionary history (Cabej, 2018a, p. 388). So, e.g., during ontogeny, cetaceans develop but later eliminate hind limb buds repeating all the initial steps (cell differentiation, formation of the apical ectodermal ridge and zone of polarizing activity, innervation, and secretion of fibroblast growth factor 8) of their terrestrial ancestors (Cabej, 2018b, Op.cit., p. 505) in a process of regressive evolution that took place between 41 and 34 million years ago

(Thewissen et al., 2006), as a developmental demonstration of their ancestral terrestrial quadrupeds. Mammal embryos still form interdigital webs as a reminder of their aquatic ancestors (Weatherbee et al., 2006).

In the first part of the book, I briefly describe the emergence and evolution of the animal multicellularity from its unicellular precursors, focusing on the morphology, reproduction, and behavior of placozoans, one of the simplest known forms of metazoan life whose origin dates back earlier than the Ediacaran eon, despite the lack of its fossil remnants apparently because of their soft body.

Chapter 2 is devoted to the Ediacaran biota whose fossils are found in Lagerstätten around the world. From the extant metazoans in the Ediacaran, only sponges left fossils, and hence a succinct description of their evolution, phylogenetic relationships, and life history will be presented, as a representative of the Ediacaran biota. The appearance of the first eumetazoans, animals with nervous system, during this period warrants a discussion on the evolution of the neuron and a description of cnidarians as a lower phylum of neuralia.

A survey of the epigenetic mechanisms of regulation of gene expression as requisites for the burst of morphological diversification of the Cambrian is presented in Chapter 3.

Chapter 4 comprises a general view of the Cambrian explosion based on the fossil record; an attempt is made to show a correlation existing between the evolutionary history of clades and emerging body plans with the evolution of the centralized bilobed brain. One focus for illustrating the transition from the diffuse to centralized nervous system (CNS) and body bilaterality will be evolution of morphology CNS and life histories of the lower metazoans, Xenacoelomorpha and Planaria.

The epigenetic control of some critical moments in the development (early development, phylotypic stage and postphylotypic development, metamorphosis, diapause, body growth), evolutionary change, transgenerational inheritance, and speciation will be presented in the last chapter.

References

Alberch, P., Blanco, M.J., 1996. Evolutionary atterns in ontogenetic transformation: from laws to regularities. Int. J. Dev. Biol. 40, 845–858.

Bromham, L., 2009. Molecular dates for the cambrian explosion: is the light at the end of tunnel an oncoming train? Palaeontol. Electron. 9 (1), 2E:3pp. http://palaeo-electronica.org/paleo/toc9_1.htm.

Budd, G.E., 2015. Early animal evolution and the origins of nervous systems. Philos. Trans. R. Soc. Lond. B Biol. Sci. 370, 20150037.

Budd, G., 2013. At the origin of animals: the revolutionary cambrian fossil record. Curr. Genom. 14, 344–354.

Butterfield, N., 2007. Macroevolution and macroecology through deep time. Palaeontology 50, 41–55.

Cabej, N.R., 2012. Epigenetic Principes of Evolution. Elsevier.

Cabej, N.R., 2018a. Epigenetic Principles of Evolution, second ed. Academic Press, London–San Diego CA, p. 388.

Cabej, N.R., 2018b. Epigenetic Principles of Evolution, second ed. Academic Press, London–San Diego CA, p. 505.

Conway Morris, S., 2000. The Cambrian "explosion": slow-fuse or megatonnage? Proc. Natl. Acad. Sci. U.S.A. 97, 4426–4429.

Darwin, C., 1859a. The Origin of Species by Means of Natural Selection, or the Preservation of Favoured Races in the Struggle for Life. John Murray, London, p. 306.

Darwin, C., 1859b. Ibid. p. 310–311.

dos Reis, M., Thawornwattana, Y., Angelis, K., Telford, M.J., Donoghue, P.C.J., Yang, Z., 2015. Uncertainty in the timing of origin of animals and the limits of precision in molecular timescales. Curr. Biol. 25, 2939–2950.

Eme, L., Sharpe, S.C., Brown, M.W., Roger, A.J., 2014. On the age of eukaryotes: evaluating evidence from fossils and molecular clocks. Cold Spring Harb. Perspect. Biol. 6, 165–180.

Erwin, D.H., Valentine, J.W., 2013. The Cambrian Explosion. Roberts and Co. Greenwood Village, Colorado, p. 16.

Kandel, E.R., 2000. The molecular biology of memory storage: a dialog between genes and synapses (Nobel Lecture). Biosci. Rep. 24, 475–522.

Linnemann, U., Ovtcharova, M., Schaltegger, U., Gärtner, A., Hautmann, M., Geyer, G., et al., 2019. New high-resolution age data from the Ediacaran–Cambrian boundary indicate rapid, ecologically driven onset of the Cambrian explosion. Terra. Nova 31, 49–58.

Maloof, A.C., Porter, S.M., Moore, J.L., Dudás, F.Ö., Bowring, S.A., Higgins, J.A., et al., 2010. The earliest Cambrian record of animals and ocean geochemical change. GSA Bulletin 122, 1731–1774.

Marshall, C.R., 2006. Explaining the cambrian "explosion" of animals. Annu. Rev. Earth Planet Sci. 34, 355–384.

Maynard Smith, J., 1990. Models of a dual inheritance system. J. Theor. Biol. 143, 41–53.

McKinney, M.L., Gittlemann, J.L., 1995. Ontogeny and phylogeny: tinkering with covariation in life history, morphology and behavior. In: McNamara, K.J. (Ed.), Evolutionary Change and Heterochrony. Wiley, Chichester, pp. 21–47.

Nichols, S.A., Dirks, W., Pearse, J.S., King, N., 2006. Early evolution of animal cell signaling and adhesion genes. Proc. Natl. Acad. Sci. U.S.A. 103, 12451–12456.

Peterson, K.J., Dietrich, M.R., McPeek, M.A., 2009. MicroRNAs and metazoan macroevolution: insights into canalization, complexity, and the Cambrian explosion. Bioessays 31, 736–747.

Stanley, S.M., 1992. Can neurons explain the Cambrian explosion? Geol. Soc. Am. Abstracts Programs 24, A45.

Thewissen, J.G.M., Cohn, M.J., Stevens, L.S., Bajpai, S., Heyning, J., Horton Jr., W.E., 2006. Developmental basis for hind-limb loss in dolphins and origin of the cetacean bodyplan. Proc. Natl. Acad. Sci. U.S.A. 103, 8414–8418.

Weatherbee, S.D., Behringer, R.R., Rasweiler, J.J., Niswander, L.A., 2006. Interdigital webbing retention in bat wings illustrates genetic changes underlying amniote limb diversification. Proc. Natl. Acad. Sci. U.S.A. 103, 15103–15107.

Further reading

Cabej, N.R., 2013. Building the Most Complex Structure on Earth. Elsevier, London - Waltham, MA.

Chapter 1

Pre-ediacaran evolution

Chapter outline

Emergence of animal multicellularity

The evolution of the prebiotic matter followed a trial-and-error pattern, whereby the second law of thermodynamics determined the stability and durability, the "correct" and faulty molecular combinations in open systems. Biotic structures are less stable, hence only temporary stable, which degrade over time. Owing to the low thermodynamic stability of proteins, the subsistence of the first living system required evolution of a mechanism of reproduction of proteins, the most versatile of the biological molecules, which was made possible with the emergence of the triplet code, leading to the evolution of the first reproductive DNA—protein system. The first living system, thus, overcame the thermodynamic barrier of life via the biological reproduction of the DNA—protein complex. The first system, DNA—protein system, combined the exceptional versatility of less stable proteins with the high thermodynamic stability of DNA capable of

1. Coding and storing in its structure the information on the associated protein (amino acid sequence) and
2. Self-reproducing the system

   This was the first informational revolution in the history of life on earth. The "fortuitous" combination DNA—protein led to the minimalist self-reproducing

Epigenetic Mechanisms of the Cambrian Explosion. https://doi.org/10.1016/B978-0-12-814311-7.00001-9

system, as a first step toward the complex protocell and cell, capable of surviving and maintaining its homeostatic state, until the system reproduced itself.

Key in the process of emergence of the living systems was the evolution of the genetic code, initially fluid, which later froze to the present standard genetic code form because of the increase in the number of proteins. The triplet code was selected as optimal for minimizing the deleterious effects of genic point mutations (Woese, 1965). The universality and optimality of the genetic code was result of a "frozen accident" (Crick, 1968) rather than consequence of stereochemical complementarity or affinities between codons and amino acids.

Thus, the first living system becomes capable of "remembering" its own structure and transmit to the offspring the information for building the same structure before it succumbed to the forces of destruction of the second law of thermodynamics.

Life on earth is believed to have emerged about 3.7 billion years ago, i.e., less than 1 billion years after Earth's formation as a planet. Fossils of microscopic bacteria indicate that these microorganisms dominated the pre-Phanerozoic multicellular animal life until ~650 Ma. These fossils also indicate that these microorganisms have been similar to the extant cyanobacteria, both morphologically and physiologically: they were anaerobic, aerobic, or facultative that reproduced asexually. By surviving without giving rise to any daughter lineages for more than 2 billion years (Butterfield, 2007), they represent a unique example of evolutionary stasis, illustrating the Simpson's concept of the "rule of the survival of the relatively unspecialized." According to recent estimates, it took more than another billion years, somewhere in the range of ca. 1007 Ma and 1898 Ma (Eme et al., 2014), for last eukaryotic common ancestor (LECA) to evolve and almost 3 billion years for multicellular animals to evolve.

Emergence of multicellularity

The first 3 billion years of evolution of life on earth may be considered as a long period of an evolutionary stasis, with the advent of eukaryotic cells about 2 billion years ago as the most important evolutionary event. More than 600 million years ago, choanoflagellates, sponges, and eumetazoans shared a common unicellular ancestor and the protostome–deuterostome ancestor (PDA) evolved −570 Ma (Peterson and Butterfield, 2005). Based on the conserved regulatory data in both protostomes and deuterostomes, Erwin believes that the PDA was in possession of the genetic toolkit of extant lower metazoans, but it had no central nervous system, eyes, appendages, segmentation, dorsal–ventral and anterior–posterior differentiation, etc. However, the "major elements of the developmental machinery required to produce sophisticated animals were already present in the PDA, even if their potential was not yet fully realized" and although "the potential for the Cambrian radiation may have been inherent in these developmental novelties, these did not drive the radiation" (Erwin, 2005). Multicellularity arose several times in the evolution of the living world (Parfrey and Lahr, 2013).

How did multicellular structure and form evolve from unicellular systems?

Metazoan morphology results not from random aggregation of cells of various types, but from specific spatial arrangement of those cells that make possible the functioning of the metazoan organism and each cell of the multicellular organism and its survival under particular environments.

For five-sixth of its existence, the life on earth evolved only slowly mainly in the form of unicellular organisms, in accordance with the Darwinian principle of gradualism. Unicellular eukaryotes emerged about 1000 million years ago, but for ~300–400 million years, they could not evolve into multicellular animals (metazoans), although the study of modern unicellular eukaryotes shows that they might have been in possession of the metazoan developmental machinery and "genetic toolkit" (Erwin, 2005).

The exceptional surge of the morphological diversity of animal forms during the Cambrian explosion required investment of huge amount of new information for erecting the complex multicellular structure, which was not present in unicellulars. Given that the genetic toolkit existed in the pre-Cambrian eon and no relevant increase in the amount or quality of genetic information did occur within that geological "blink of an eye," from the Darwinian standpoint, Cambrian explosion is an unpredicted and paradoxical event. Biology must answer the question: Where did that huge amount of nongenetic information come from? An attempt will be made in this section to explain the source of the epigenetic information that made possible the Cambrian explosion.

Henry Quastler's postulate that "creation of new information is habitually associated with conscious activity" (Quastler, 1964) is too restrictive. From the current status of knowledge, we could affirm that generation of new information in eumetazoans is a property of the neural activity. But this is challenged by the fact that metazoans emerged as nerveless multicellular structures, in all likelihood resembling extant placozoans and sponges.

Choanoflagellates

Based on the hypothesis that multicellular organisms evolved from unicellulars, biologists began their search for the origin of multicellular animals, metazoans, by focusing on choanoflagellates for a good reason; choanoflagellates resemble choanocytes, a particular type of sponge cells. No choanoflagellate fossils have been found yet.

Choanoflagellates, sponges, and other metazoans share a common unicellular ancestor living more than 600 million years ago. Geneticists have sequenced the genome of a free-living choanoflagellate, *Monosiga brevicollis* (King et al., 2008), which consists of 9196 genes. Most of these genes derive from the common ancestor of choanoflagellates, placozoans, and eumetazoans. The size of the genome is considerable for a unicellular organism. It contains many families of genes for eukaryotic transcription factors, including two homeodomain proteins, previously thought to be specific to metazoans. *M. brevicolli's* genome also contains protein domains associated with cell adhesion in metazoans as well as a surprising number of tyrosine kinase and their downstream elements. All of them seem to have been inherited from the unicellular common ancestor of choanoflagellates and metazoans: "The common ancestor of metazoans and choanoflagellates possessed several of the critical structural components used for multicellularity in modern metazoans" (King et al., 2008). The above facts remind us of G.G. Simpson's (1902—84) idea: "All the essential problems of living organisms are already solved in the one-celled … protozoan and these are only elaborated in man or the other multicellular animals" (Brunet and Arendt, 2016).

Choanoflagellates also possess various synaptic protein homologs, and a neurotransmitter secretion mechanism is conserved in both choanoflagellates and metazoans (Burkhardt, 2015). *M. brevicollis* has a substantial set of cell adhesion proteins, many tyrosine kinases, and their downstream signaling targets, but lacks most of intercellular signaling pathways and a number of transcription factors of the metazoan developmental toolkit (King et al., 2008). It also lacks other metazoan genomic features (e.g., several genes for metazoan-specific signaling pathways), but this may be result of the loss in *M. brevicollis* in the course of its phylogeny or, even more likely, that these features were later added to the genetic toolkit of metazoans.

Transition to multicellularity: new structure, new rules of game

The control system of unicellulars lacks the information necessary for determining metazoan Bauplan and morphology, the specific arrangement of different types of cells in distinct groups to form particular tissues and organs. The challenge to building a multicellular structure was evolution of a novel control system that would be capable of coordinating the function and behavior of tens to billions of cells of different types as well as memorizing and transmitting to the offspring instructions for building the species-specific structure.

Metazoans may not have evolved earlier than ~650 Ma (Morris, 1998). From an evolutionary standpoint, the emergence of multicellular animals was an extremely difficult task. Cells in a multicellular organism do not simply obey forces of interaction of their own components, but they also have to adjust the form, function, and behavior to the requirements of the organism. This implies that each cell in a multicellular system is subordinate to the control system of the organism. Individual cells in the multicellular organism are no longer independent living systems; they live a double life.

Unlike unicellulars that are self-sufficient and live a biologically independent life, somatic cells in a metazoan organism have to relinquish the ability for free independent life, as well as lose their pluri/totipotency and complexity at the morphological and functional levels. Living in the wondrous company of billion/trillion cell large cell community is not cost-free and they have to live a double life for themselves by performing functions related to sustaining their own life and for the organism by contributing to maintenance of its functions and structure.

The multicellular life evolved from unicellular organisms and similarly a multicellular organism in the process of development also begins from a single cell, ovum or zygote, which, via successive divisions, produce cells that progressively lose their pluri/totipotency to specialize in performing specific physiological and/or structural functions, leading to formation of different types of cells. However, a common property of all cell types is that they use the same basic pattern of expression of housekeeping (tissue-specific) genes for sustaining their own life and, additionally, establish their cell type—specific patterns of expression of tissue-specific genes for performing functions at the organismic, supracellular, and systemic levels.

Evolution of the control system

Multicellularity evolved independently in at least 16 different eukaryotic lineages, including animals, plants, and fungi (Brunet and King, 2017), suggesting that their ancestors possessed the genetic toolkit determining the Bauplan.

Metazoans are physically improbable structures and maintenance of their improbable information-consuming structures against the thermodynamic forces of disorder and destruction required evolution of a *control system*, or they would quickly succumb to those forces. Temporary thwarting of the effects of these forces is the *raison d'être* of the control system in living beings.

Unicellulars could not have "predicted" the advent of multicellular life to preemptively evolve mechanisms for determining the structure and coordinating the activity of various cell types for erecting the multicellular structure. The unicellular genetic toolkit and genes, however indispensable for cell differentiation and building the metazoan structure, lack the information on the animal structure, i.e., the specific arrangement of the myriad of cells of tens to hundreds of different types.

Cell division, differentiation, and growth are results of controlled processes that lead to species-specific arrangement of various cell types in formation of the Bauplan, tissues, and organs. A special role in determining the direction of growth in the embryo plays the orientation of the mitotic spindle, a microtubular structure consisting of three types of MTs (microtubules): interpolar MTs extending between the two cell poles, astral MTs, connecting them to the cell cortex and kinetochores, which attach to chromosomes to appropriately allocate them in two daughter cells (Anderson et al., 2016).

Whether the dividing cells will be located anteriorly/posteriorly, dorsally/ventrally, or laterally is determined by the position of the mitotic spindle relative to the neighboring cells. The position of the mitotic spindle, in turn, depends on a scaffolding protein complex that links a cortical molecule to the astral MTs of the spindle. The scaffold is the GK_{PID} (glycerol kinase protein interaction domain) of the Dlg (Discs large), which binds MT-associated motor proteins, KHC-73 and Pins (Partner of inscuteable protein) in insects and LGN (leucine—glycine—asparagine tripeptide containing protein) in vertebrates (Fig. 1.1).

It seems that Dlg's GK_{PID} evolved from a gk (glucokinase) enzyme. Choanoflagellates possess both gk enzyme and GK_{PIDs}. The latter evolved via the duplication of the *gk* gene. The GK_{PID} acquired the ability to bind Pins and determine spindle orientation before the evolution of metazoans and multicellularity.

In the choanoflagellate—metazoan ancestor, the spindle orientation was determined by the position of flagellar basal bodies, which enables formation of only spherical colonies. The advent of the GK_{PID}-Pins coupling deprived

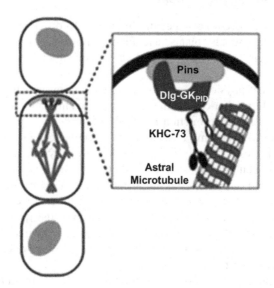

FIGURE 1.1 Function and phylogeny of the guanylate kinase (gk) and GKPID protein family. The GK_{PID} of the protein Discs-large (Dlg, blue) (dark gray in orint version) serves as a scaffold for spindle orientation by physically linking the localized cortical protein Pins (green) (gray in print version) to astral microtubules (red) (light gray in print version)via the motor protein KHC-73 (black) (black in print version). *From Anderson, D.P., Whitney, D.S., Hanson-Smith, V., Woznica, A., Campodonico-Burnett, W., Volkman, B.F., et al., 2016. Evolution of an ancient protein function involved in organized multicellularity in animals. eLife 5 (2016), e10147.*

cells of the enzyme activity and provided them with a mechanism to orient the spindle toward the adjacent cells via an externally organized molecular mark. For the inhibition of the enzyme activity and refunctionalization of GK_{PID} activity relative to the spindle orientation to occur, only one of the two substitutions (s36P or f 33S) was necessary to occur in metazoans.

The evolution of GK_{PIDS} was a gradual process of molecular exploitation (Fig. 1.2), in which the preexisting molecules were recruited by newly evolved molecules to perform a new function (Anderson et al. (2016)).

The choanocyte genome

Recently, biologists have sequenced the genome of a free-living choano-flagellate, *M. brevicollis* (King et al., 2008), which consists of 9196 genes, which is unusually large for a unicellular organism. Most of these genes certainly derive from the common ancestor of choanoflagellates, placozoans, and eumetazoans in the late Precambrian period, more than 600 million years ago (King et al., 2008). Choanoflagellates possess a genetic toolkit that comprises genes for many families of metazoan transcription factors, cell signaling, cell adhesion, signaling pathways, *Hox* genes, etc., and a surprising number of tyrosine kinases and their downstream elements. No other known unicellular eukaryote has any of the metazoan developmental signaling pathways (King et al., 2008). It contains two homeodomain proteins, previously thought to be specific to metazoans. *M. brevicollis* genome also contains protein domains associated with cell adhesion in metazoans. All these genes seem to have been inherited from the unicellular common ancestor of choanoflagellates and metazoans. "The common ancestor of metazoans and choanoflagellates possessed several of the critical structural components used for multicellularity in modern metazoans" (King et al., 2008).

The fact that many metazoan genomic features (e.g., genes for metazoan-specific signaling pathways) are absent in the choanoflagellate is believed to be result of the gene loss in *M. brevicollis* in the course of its evolution from the common ancestor of choanoflagellates and metazoans, but it seems more likely that these features were later added to the genetic toolkit of metazoans.

Choanoflagellates as precursors of metazoan multicellularity

A basic tenet of the evolutionary biology is that multicellular life evolved from unicellulars. The search for possible unicellular ancestors of the first metazoans led biologists to choanoflagellates, whose resemblance to sponge choanocytes was noticed over 150 years ago. Now choanoflagellates are considered more closely related to metazoans than any other known unicellular animal. They show a very close resemblance to sponge choanocytes (Brunet and King, 2017; King et al., 2008) (Fig. 1.3).

Protein complex evolution   Organized tissue evolution

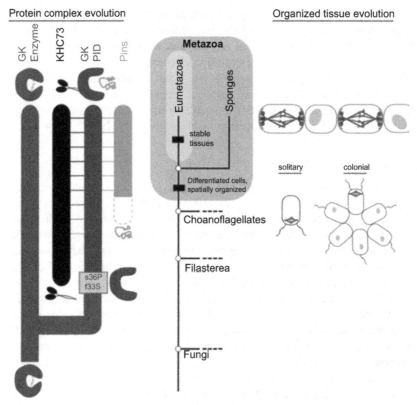

FIGURE 1.2 **Historical evolution of GKPID-mediated spindle orientation complex.** The center portion shows the phylogeny of Metazoa and closely related taxa. The origin of cell differentiation and spatially organized tissues is marked. The left portion shows major events in the evolution of the components of the spindle orientation complex reconstructed in this study. Duplication of an ancestral gk enzyme (brown) (gray in print version) and the key mutations that led to the origin of a GK_{PID} (blue) (dark gray in print version) that could bind other molecules in the complex are shown relative to the phylogeny's time scale. The apparent date of origin of KHC-73 (black) (black in print version) and Pins (green) (light gray in print version) is also shown. Dotted green line shows the origin of Pins in a form not yet bound by GK_{PID}. Solid green line shows GK_{PID}-binding form. Horizontal lines indicate binding between proteins. The right portion shows a schematic of the spindle orientation machinery in metazoans, which allows orientation relative to external cues from nearby cells, as well as spindle orientation relative to the internal cell axis as marked by the flagella in both solitary and colonial choanoflagellates. *From Anderson, D.P., Whitney, D.S., Hanson-Smith, V., Woznica, A., Campodonico-Burnett, W., Volkman, B.F., et al., 2016. Evolution of an ancient protein function involved in organized multicellularity in animals. eLife 5 (2016), e10147.*

Choanoflagellates are not derived from metazoans; as a separate lineage, they evolved before the emergence of metazoans (King et al., 2008).

Choanoflagellates form colonies of several cells. The development of colonies in this group results from reproduction of similar cells (Figs. 1.4 and

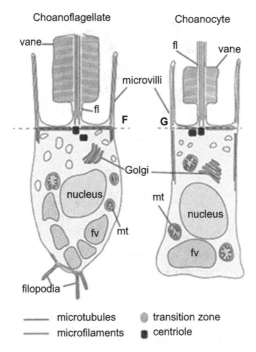

FIGURE 1.3 Comparative ultrastructural schematics of a choanoflagellate and a sponge choanocyte, modified from *Maldonado, M., 2004. Choanoflagellates, choanocytes, and animal multicellularity. Invertebr. Biol. 123, 1–22*, following (Woollacott and Pinto, 1995) for the microtubule cytoskeleton and (Karpov and Coupe, 1998) for the actin cytoskeleton. *mt*, mitochondria; *fl*, flagellum; *fv*, food vacuole. *From Brunet, T., King, N., 2017. The origin of animal multicellularity and cell differentiation. Dev. Cell 43, 124–140.*

1.5) rather than from aggregation of similar cells as demonstrated by the experimental evidence that the use of the cell cycle inhibitor aphidicolin prevents formation of choanocyte colonies. There is no evidence that choanoflagellates may form colonies via aggregation of similar cells (Fairclough et al., 2010).

A characteristic of all multicellular organisms, beginning from the simplest ones, is the presence of epithelium, a one-cell thick tissue that separates and protects underlying tissues from the harmful environmental agents. It also lines the inner cavities and free surfaces of tubular organs. Formation of epithelia required emergence of a stable mechanism of cell adhesion that seems to have independently evolved in several groups of living forms and connect epithelial cells with each other and with the intercellular matrix. The last common ancestor of metazoans is inferred to have been in possession of various intercell adhesion junctions (Abedin and King, 2010) (Fig. 1.6).

Unlike choanoflagellates, where every cell can feed on their own, most of choanocytes in sponges lost ability to feed (Cavalier-Smith, 2017). In

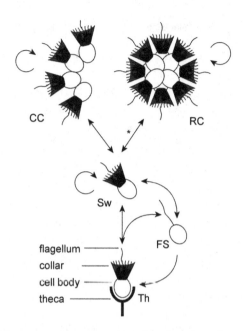

FIGURE 1.4 *Salpingoeca rosetta* can transition through at least five morphologically and behaviorally differentiated cell types (Fairclough et al., 2013). Solitary "thecate" cells attached to a substrate (Th) can produce solitary swimming (Sw) cells or solitary fast swimming (FS) cells, either through cell division or theca abandonment. Solitary swimming cells can divide completely to produce solitary daughter cells or remain attached after undergoing incomplete cytokinesis to produce either chain colonies (CC) or rosette colonies (RC) in the presence of the bacterium *Algoriphagus machipongonensis* (asterisk). *Fil*, Filasterea; *Cho*, Choanoflagellates. *From Fairclough, S.R., Chen, Z., Kramer, E., Zeng, Q., Young, S., Robertson, H.M., Begovic, E., et al., 2013. Premetazoan genome evolution and the regulation of cell differentiation in the choano-flagellate Salpingoeca rosetta. Genome Biol. 14 (2013), R15.*

Cavalier-Smith's view, the choanocyte swimming ball colony could evolve via somatic cell differentiation into a basal pinacocyte that secretes collagen and mucopolysaccharide-rich extracellular matrix and form a mesohyl layer with spicules and rare cells between the external pinacocytes and internal choanocytes. This new layer could serve as mesohyl skeleton between two layers and support a larger presponge structure that could anchor itself on a rock. The same genetic tool used for differentiating pinacocytes could be used for gamete (sperm cells and larger oocytes) differentiation (Cavalier-Smith, 2017).

It is not resolved yet, which of the nonbilaterian phyla is the sister group of the rest of metazoans but here will be preferred to consider placozoans, in line with other authors (Srivastava et al., 2008; Schierwater et al., 2009; Schlei-cherová et al., 2017; Eitel et al., 2018).

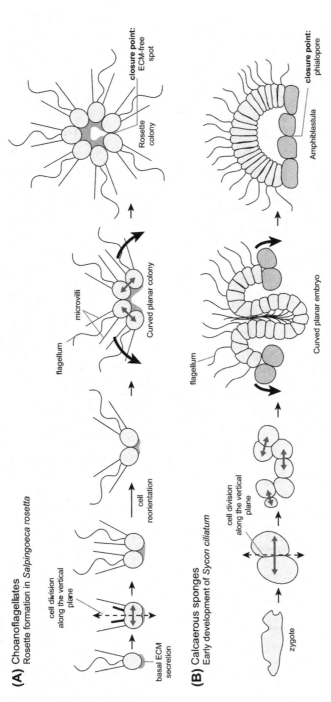

(A) Choanoflagellates
Rosette formation in *Salpingoeca rosetta*

cell division
along the vertical
plane

basal ECM
secretion

cell
reorientation

flagellum

microvilli

Curved planar colony

closure point:
ECM-free
spot

Rosette
colony

(B) Calcaerous sponges
Early embryonic development of *Sycon ciliatum*

zygote

cell division
along the vertical
plane

flagellum

Curved planar embryo

closure point:
phialopore

Amphiblastula

FIGURE 1.5 Calcareous sponge embryos and Volvocale embryos. (A) Morphogenesis during rosette formation in the choanoflagellate *Salpingoeca rosetta*, following Fairclough et al. (2010). (B) Early embryonic development of the calcareous sponge *Sycon ciliatum*, including amphiblastula inversion *(from Franzen, W., 1988. Oogenesis and larval development of Scypha ciliata (Porifera, Calcarea). Zoomorphology. 107, 349–357). From Brunet, T., King, N., 2017. The origin of animal multicellularity and cell differentiation. Dev. Cell 43, 124–140.*

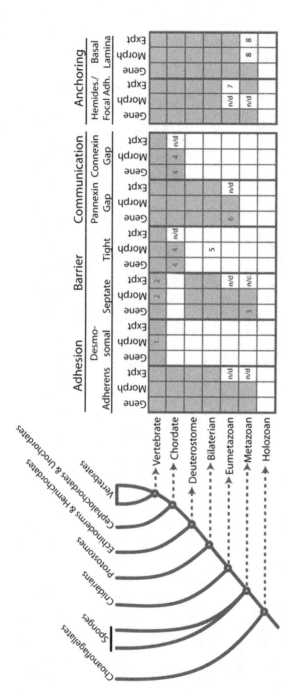

FIGURE 1.6 Evolutionary history of epithelial cell junctions. The four main functions of animal epithelia are adhesion, barrier, signaling, and anchoring. Elements of adherens, septate, pannexin-based gap, and anchoring junctions (hemidesmosomes [Hemides.], focal adhesion [Focal Adh.], and basal lamina) were likely present in the last common ancestor of animals, while desmosomes, tight, and connexin-based gap junctions emerged later in animal evolution. For a series of ancestors representing stages in animal evolution, the presence (filled box) or absence (open box) of evidence for (1) genes diagnostic of each junction type ("Gene"), (2) junction morphology detected by electron microscopy ("Morph"), and (3) experimental support (e.g., stable cell adhesion, barrier function, or protein localization; "Expt") in extant lineages is indicated (Tyler, 2003; Boury-Esnault et al., 2003; Sakarya et al., 2007; Abedin and King, 2008; Spiegel and Peles, 2002; Bing-Yu et al., 1997; Banerjee et al., 2006; Nichols et al., 2006; Ledger, 1975; Leys et al., 2009; Shestopalov and Panchin, 2008; Chapman et al., 2010; Litvin et al., 2006; Boute et al., 1996; Ereskovsky et al., 2009; Aouacheria et al., 2006; Muller and Muller, 2003; Brower et al., 1997; Magie and Martindale, 2008; Sabella et al., 2007). *From Abedin, M., King, N., 2010. Diverse evolutionary paths to cell adhesion. Trends in cell biology 20, 734–742.*

An "epigenetics first" hypothesis of emergence of multicellularity

During the last two decades, S.A. Newman, G.B. Müller, R. Bhat, and others came up with an "epigenetics comes first" hypothesis, which posits that the evolution of incipient multicellular organisms was facilitated and promoted by the physical morphogenetic and patterning effects. Accordingly, formation and stability of the early pre-Cambrian multicellulars required as a sine *qua non* the presence of cell-to-cell adhesion molecules, proteins, and a few polysaccharides, which evolved earlier in many unicellulars, including *M. brevicollis*, a choanoflagellate, the closest unicellular relative, if not precursor, of the first metazoans. Choanoflagellates and other unicellulars have on their surface a diversity of cadherin homologs and a few adhesive polysaccharides, which bind together cells in multicellular aggregates (colonies or primitive multicellular animals), like the cells after the zygotic division (Niklas and Newman, 2013).

The presence of cadherins, a number of toolkit genes, and cell-to-cell adhesive compounds produced during the cell aggregation, along the physical effects of the aggregation and adhesion, determine the shape and pattern, i.e., the "specific arrangement of cell types" of the multicellular systems (Fig. 1.7), hence representing dynamical patterning modules (DPMs) (Newman and Bhat, 2009). Many DPMs appeared first in choanoflagellates, although in unicellulars they perform other functions.

Unlike colonies of the choanoflagellate *Salpingoeca rosetta*, and maybe of *Capsaspora owczarzaki*, two of the closest relatives of metazoans, whose cells are firmly attached via cytoplasmic bridges, the cadherin-based adhesion in animal embryos, forms "liquid tissues," which become immiscible by spatial modulation of cadherins. A "repurposing" of ancestral cadherins via their linkage to the cytoskeleton and Wnt expression in unicellular holozoans might have opened the way for emergence of diploblastic eumetazoans.

The first group of DPMs activated in primitive multicellulars has been that of cell adhesion. At this early stage, the connection between the phenotype and the genotype must have been loose, hence authors describe this stage as a "pre-Mendelian world" (Newman and Müller, 2001a). Cells in these multicellular aggregates were inclined to sort out "into islands of more cohesive cells within lakes composed of their less cohesive neighbors," giving rise to separate cell layers (Newman and Muller, 2001). The part of the motile nonadhesive epithelioid cells would readily develop cavities resembling those observed in cnidarians. Accordingly, lumen formation was result of differential cell adhesion. Other products of these epigenetic mechanisms included tubes, lobes, segmentation, as well as multilayered and appendage-like structures (Newman and Müller, 2001b; Newman and Bhat, 2009). These novelties then are fixed in ontogeny, enabling novelties to develop independently of the perturbations by external factors, such as temperature and pressure (Love et al., 2017).

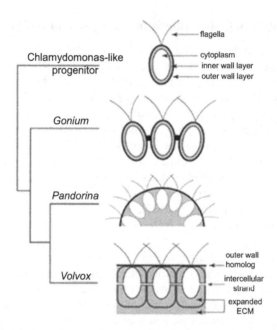

FIGURE 1.7   Putative evolution of multicellularity in the volvocine algae (*adapted from Kirk, D.L., 2005. A twelve-step program for evolving multicellularity and a division of labor. BioEssays 27, 299–310*). The inner cell wall layer of a unicellular *Chlamydomonas*-like progenitor is modified into an expanded extracellular matrix (ECM) wherein multicellularity is achieved by intercellular cytoplasmic strands (e.g., *Volvox*). The outer cell wall layer adheres cells in colonial organisms (e.g., *Gonium*). From Niklas, K.J., Newman, S.A., 2013. *The origins of multicellular organisms. Evol. Dev. 15, 41–52.*

According to the hypothesis, epigenetic processes based on DPMs were the motive forces that determined the early evolution of the animal morphology and prevailed in the initial stages of the pre-Cambrian evolution. Over time, genetic changes that tended to conserve the epigenetically determined phenotypes were coopted, thus transforming them into genetically heritable phenotypes in a process of genetic assimilation. Stepwise, the initial "one-to-many" genotype–phenotype relationship evolved into the present-day "one-to-one" relationship (Newman and Muller, 2001). Under the action of natural selection, genetic changes that happened to favor the maintenance of stable morphological characters coopted those phenotypes making them heritable.

Thus, the contingent and conditional epigenetic processes were the motive forces of evolution at the dawn of the metazoan life. In any case the "pre-Mendelian organisms" had no Bauplan and were morphologically fluid, with no fixed or characteristic shape that characterize Cambrian metazoans.

Placozoans

Placozoans (from ancient Greek zoo (ζώο) "animal" and plax (πλάξ) "plate"), along sponges, are the most primitive of extant metazoan groups (Philippe et al., 2009) or even older (Schleicherová et al., 2017; Eitel et al., 2018) than sponges and may be considered "living fossils" (Schierwater et al., 2009).

Placozoans are small (1–2 mm), flat, and thin benthic animals, consisting of two (upper and lower) epithelial layers and a layer of loose fiber cells between them. They show no Bauplan or body symmetry. Like sponges, they have no nervous system, but they have only four to six cell types, i.e., half the cell types sponges have (choanocytes, amoebocytes, pinacocytes, porocytes, sclerocytes, myocytes, oocytes, sperm cells, lophocytes, collencytes, rhabdiferous cells). They have no nerve cell/nervous system, which is a general trademark of eumetazoans, and no digestive, circulatory, or excretory systems.

Their study may be crucial to understanding the early stages and minimal requirements of eumetazoan organization (Schierwater, 2005). Until quite recently, *Trichoplax adhaerens* (Schulze 1883) was considered to be the only extant representative of the group, but in 2018 a new species, *Hoilungia hongkongensis*, nov. spec., Eitel, Schierwater, and Wörheide, was found and added to the group (Eitel et al., 2018).

Morphology

Placozoans are millimeter-size thin metazoans, consisting of two (upper and lower) epithelial layers and a layer of loose fiber cells between them. With the lower epithelial layer, they attach to the substrate and glide over it. Placozoans have only four to six cell types (ciliated dorsal and ventral epithelial cells; fiber cells, located between the epithelial cell layers; gland cells, resembling neurons and neurosecretory cells; ventral lipophilic cells secreting alga-digesting enzymes; crystal cells) (Fig. 1.8), but, like sponges, they lack muscle and sensory cells, nervous and digestive, circulatory, and excretory systems. They feed through external digestion at the ventral surface (Smith et al., 2015).

Reproduction

Commonly, placozoans reproduce asexually by binary fission, but reproduction by budding is also observed. Sexual reproduction is demonstrated by production of eggs and their development, but it occurs only at temperatures 23°C and higher, perhaps under unfavorable conditions, such as higher population density and food depletion (Eitel et al., 2011). After producing egg(s), the mother dies and degenerates (Fig. 1.9).

The embryonic development, however, has not been possible to follow because, under laboratory conditions, it stops at 128 cell stage (Eitel et al., 2011). Five sperm markers of various stages of spermatogenesis are identified

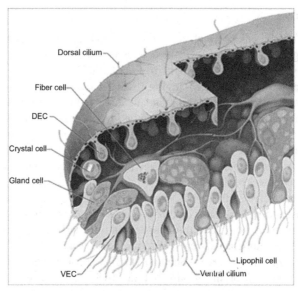

FIGURE 1.8 Drawing summarizing *Trichoplax* cell types and body plan. Facing the substrate (below) is a thick ventral plate composed of ventral epithelial cells (VEC; light yellow), each bearing a cilium and multiple microvilli; lipophil cells (brick) that contain large lipophilic inclusions, including a very large spherical inclusion near the ventral surface (lavender); and gland cells (pale green), distinguished by their contents of secretory granules and prevalence near the margin. Dorsal epithelial cells (DEC; tan) form a roof across the top from which are suspended their cell bodies surrounded by a fluid-filled space. In between the dorsal epithelium and ventral plate are fiber cells with branching processes that contact each of the other cell types. A crystal cell (pale blue) containing a birefringent crystal lies under the dorsal epithelium near the rim. *From Smith, C.L., Varoqueaux, F., Kittelmann, M., Azzam, R.N., Cooper, B., Winters, C.A., et al., 2014. Novel cell types, neurosecretory cells and body plan of the early-diverging metazoan, Trichoplax adhaerens. Curr. Biol. 24, 1565−1572.*

in *Trichoplax* and nonflagellated round cells in the intermediate suspected of being sperm are described, but fertilization has never been observed (Miller and Ball, 2005). The sexual reproduction in the *T. adhaerens* is poorly known, but genetic studies have shown that it has the genetic tools that cnidarians and bilaterians use to segregate the germ line (Srivastava et al., 2008).

Empirical evidence suggests that differentiation and maturation of sperm and egg cells takes place in different individuals, implying that *Trichoplax* is a bisexual organism. Molecular signatures of sexual reproduction (low nucleotide polymorphism, intergenic or interchromosomal recombination, and sharing of alleles between heterozygotes and homozygotes) are recorded (Signorovitch et al., 2005; Eitel et al., 2011; Eitel et al., 2018).

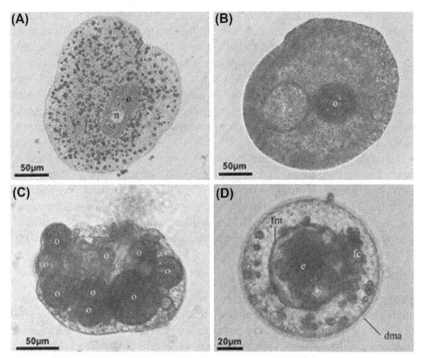

FIGURE 1.9 Progress of *Placozoa* sp. H2 oocyte maturation and early embryogenesis. Shown are light microscopy (A–D) images of the putative *Placozoa* sp. H2 oocytes and embryos. Typically, one oocyte with a large nucleus starts growing in a flat animal without any signs of degeneration (A). Accompanied by the generation of yolk droplets (B), the animal enters the degeneration phase (D-phase) after 5–6 weeks when the population density reaches its maximum. Occasionally, several oocytes are found in a single degenerating animal. We found one animal with nine maturing oocytes (C). The D-phase starts with the lifting of the upper epithelium followed by condensing the lower epithelium until forming a hollow sphere ("brood chamber") containing the embryo (D). The oocyte grows until reaching a varying final size of 50–120 μm by incorporating extensions from fiber cells through pores. *n*, nucleus; *o*, oocyte; *yo*, yolk outside oocyte; *fm*, fertilization membrane; *e*, embryo; *fc*, fiber cells; *dma*, degenerating mother animal. *From Eitel, M., Guidi, L., Hadrys, H., Balsamo, M., Schierwater, B., 2011. New insights into placozoan sexual reproduction and development. PLoS One 6(5), e19639.*

The genome

The placozoan genome consists of six chromosome pairs harboring a complement of ~15,000 genes (Srivastava et al., 2008), of which 80% are shared with cnidarians and bilaterians. This is equal to the ~15,000 loci found in a eumetazoan like planaria (Adamidi et al., 2011) and comparable to ~20,000 genes of the human genome. There is evidence indicating that different placozoan lineages are genetically as different as representatives of different families in other metazoan phyla (Voigt et al., 2004), what led to the idea of existence of different putative species in the group and the classification of *H. hongkongensis* considered to belong to a new genus because of the differences observed in its genome compared to *T. adhaerens.*

Placozoan gene complement and structure is highly conserved and, surprisingly, the arrangement of genes in its genome is similar to the one observed in the human genome. This is in stark contrast with the loss of ancestral arrangement of genes in the genomes of nematodes and fruit flies (Srivastava et al., 2008).

Contrary to what one would expect, placozoans, probably the simplest of extant known metazoans, have a great repertoire of transcription factors: 35 homeobox genes, 14 ANTP-class genes, 9 PRD-class (paired box and homeobox), 27 helix−loop−helix, 56 zinc finger, etc. The *Trichoplax* genome contains sequences for an FMRFamide-like neuropeptide, two additional neuropeptides, and several classical neurotransmitters (Srivastava et al., 2008).

Behavior

Although placozoans have no nervous system, they perform various typical animal behaviors, such as locomotion, feeding, phototaxis, and chemotaxis. Genomic studies show that they have the basic set of genes for performing functions of the nervous system (Srivastava et al., 2008; Varoqueaux and Fasshauer, 2017). Besides, nonneural organisms such as placozoans and Porifera also sense changes in the environment and receive external stimuli (light, chemicals, food), as indicated by the fact that they respond adaptively to them, by changing their behavior (Ludemanet al., 2014). Recently, biologists observed that in the absence of food, *Trichoplax* move by gliding, rotating in place, and folding, but when they come in contact with algal food, they stop moving and beating cilia, lyse underlying algae, and begin digesting the lysate, in a process of external digestion that is unique in the Kingdom Animalia (Smith et al., 2015), indicating the presence of a sensory system linked to effectors that control cilia beating throughout the ventral epithelium and the whole body (Senatore et al., 2017).

Arrest of the ciliary beating and gliding, like the one observed during feeding, was also elicited by administration of some endomorphin-like peptides (ELPs) that are similar to opioid neurotransmitters. In this case, the arrest of ciliary beating in one animal induces ciliary arrest in the neighboring animal and so did the latter to its neighboring animals, successively. It was also observed that when the animal paused in the absence of food, the pausing behavior spread progressively to other animals over distances of about several millimeters, indicating that same diffusible factor was released by each affected animal. And since the pausing behavior in this case was identical with the one elicited by ELPs, it was inferred that the diffusible released by the ventral epithelial secretory cells may be an ELP chemical signal. The ELP secretory cells, by releasing into the environment the neuropeptide, stimulate neighboring secretory cells to do the same (Senatore et al., 2017). The fact that a small number of ventral secretory cells may induce ciliary arrest across the entire animal body suggests that these cells somehow, in the absence of nervous system and synaptic contacts, control the ciliary beating in effector cells (Senatore et al., 2017). The observed modulation of the ventral epithelial cells by the presence or absence of the food indicates that they are chemosensory cells (Smith et al., 2015) that control the feeding behavior and the external digestion.

Ciliated secretory cells are common in ctenophores, cnidarians, bilaterians (Senatore et al., 2017) and also in sponges' larvae where they "convert environmental cues into internal signals via Ca^{2+}-mediated signaling, which is necessary for the initiation of metamorphosis" (Nakanishi et al., 2015).

Voltage-gated calcium (C_{av}) channels are essential for the function of neurons (Clapham, 2007). Of the three types of C_{av} channels ($C_{av}1$, $C_{av}2$, and $C_{av}3$) involved in translation of electrical stimuli in neurons and muscles of bilaterians, a single $C_{av}3$ ($TC_{av}3$) (Trichoplax $C_{av}3$), which is similar to mammalian $C_{av}3$, and its structure is well-conserved in the gland cells of *T. adhaerens*, despite the very long time of more than 600 million years of divergence from other C_{av} channels. Based on the homology of other placozoan cell types with the corresponding bilaterian cells and the biophysical properties of $TC_{av}3$, it is predicted to play a role in the excitability of the gland cells, "which resemble neurons and neurosecretory cells in their expression of exocytotic SNARE proteins and membrane-apposed vesicles" (Smith et al., 2017).

It is suggested that the ciliated lower epithelial cells of placozoans may also use Ca^{2+}-mediated signaling in their behavioral responses (feeding and motile behaviors such as phototaxis and chemotaxis) to environmental stimuli. Indeed, the presence of electrogenic genes and $C_{av}3$ channels indicates that electrical and Ca^{2+} signaling occur in placozoans (Senatore et al., 2016). Placozoans also have several neurotransmitters (Srivastava et al., 2008).

The control system in placozoans

Placozoan adaptive responses to detecting food, light, and chemicals (arrest of movement and external digestion, phototaxis, and chemotaxis) imply perception of the external and internal environments and require coordination of the activity of numerous cells. All these and their response to deterioration of the conditions in the environment (depletion of food resources and increased population density) by starting production of ova and the process of reproduction indicate the existence of a still unknown and understudied control system.

The role of the control system (schematized in Introduction Fig. 1.1) in eumetazoans is performed by the nervous system. What plays the role of the control system in these aneural animals? We have some important hints, but we are still far from a real knowledge of their control system, so the only possible approach at the present time is to try to logically put together the known pieces of puzzle of placozoan behavior control.

If the fiber cells, which make contacts with almost all the cells of the body, serve to spread the Ca^{2+}-mediated signaling across the entire placozoan body, this would provide a parsimonious explanation of the coordination of the cilia beats during locomotion related to placozoan feeding, phototaxis, and chemotaxis.

With six or more processes and more secondary branches, fiber cells make contacts with the ventral and dorsal epithelial cells and lipophilic cells, as well as between each other, reaching almost all the cells of the organism (Smith et al., 2014) (Fig. 1.10). This integrative morphology of fiber cells is reminiscent of the pervasive presence of the nervous system in neural animals that via nerve endings reach all the cells of the organism. Fiber cells do not form synapses, but processes extending between them have electron dense septa, which while preventing large molecules and organelles to cross the junction allow communication between cell plasmata, thus forming an uninterrupted syncytium of all fiber cells in the placozoans organism. Like plug junctions in Hexactinellid sponges, they allow intercellular ion flow, thus enabling propagation of electrical impulses. It is suggested that these septa are forerunners of synapses (MacKie and Singla, 1983).

FIGURE 1.10 Scanning electron microscope image of interior surface exposed by removing the dorsal epithelium. Regularly spaced fiber cells (F) extend multiple tapering processes around the cell bodies of underlying ventral epithelial cells (v). Lipophil (L) is identified by its content of granules. Inset: Transmission electron microscopy shows dense septum bisecting the cytoplasm of a process resembling that of a fiber cell. The membrane is continuous across the septum. *From Smith, C.L., Varoqueaux, F., Kittelmann, M., Azzam, R.N., Cooper, B., Winters, C.A., et al., 2014. Novel cell types, neurosecretory cells and body plan of the early-diverging metazoan, Trichoplax adhaerens. Curr. Biol. 24, 1565–1572.*

It is suggested that gland cells may function as neurosecretory cells (Schierwater et al., 2008). Senatore and colleagues have identified ~1500 ELP (endomorphin-like peptide) secretory cells located around the edge of the animal, which are likely to be chemosensory neurosecretory cells that on detection of algae secrete ELP in the seawater, thus inducing arrest of cilia beating and locomotion to start feeding and external digestion. The function of these placozoan cells is seen as "analogous to that of sensory neurosecretory cells and neurons that modulate the activity of ciliated cells and other types of effector cells in animals with nervous systems" (Senatore et al., 2016). Other authors have suggested that some gland cells in *T. adhaerens* may function as neurosecretory cells that in a paracrine mode control feeding and locomotor behavior (Smith et al., 2014). In addition to the previously identified secretory cells, two other cell types are described with some of them secreting ELP (Smith et al. (2014).

It is a widely accepted opinion that the precursor of the nervous system was "a system of coordinated sensory secretory cells" (Senatore et al., 2017). *T. adhaerens* has neurosecretory cells (Smith et al., 2014) producing several signaling molecules, such as FMRFamide neuropeptides, RFamide-like neuropeptides (Senatore et al., 2017), and two predicted RWamide-like regulatory peptides (Nikitin, 2014). These signaling neuropeptides have been important for coordination of the *T. adhaerens* physiology and behavior in these nerveless animals, in the initial steps of the metazoan evolution. Such chemical mechanisms of signal transmission, along the synaptic transmission of signals, are still conserved to various degrees in the brain of many extant animals (Williams et al., 2017).

Evolution

Placozoans are thought to have evolved during the Ediacaran eon (635–542 Ma) or Cryogenian era (720–635 Ma) and by some authors are considered to be the most primitive metazoans, older than sponges, and hence their study may be crucial for understanding early stages of metazoan evolution. Based on analyses of the mitochondrial DNA, investigators place placozoans "at the root of the Metazoa" (Dellaporta et al., 2006).

None of several hypotheses on the origin of placozoans has found sufficient support. One of the hypotheses, based on the similarity of their mitochondrial genome size with that of choanoflagellates and the minimal morphological organization, considers placozoans to be sister to all other animals (Dellaporta et al., 2006; Osigus et al., 2013) (Fig. 1.11).

In regard to the origin of placozoans, it is noteworthy the hypothesis that the Ediacaran genus *Dickinsonia* may be a stem placozoan. The hypothesis is based on the fact that *Dickinsonia* trace fossils indicate that it used the whole

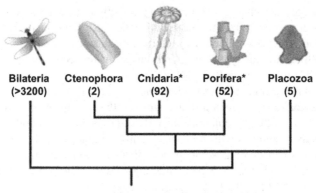

Bilateria	Ctenophora	Cnidaria*	Porifera*	Placozoa
(>3200)	(2)	(92)	(52)	(5)

* linear or circular mitochondrial genomes

FIGURE 1.11 Available mitochondrial genome data for Bilateria and the four diploblastic metazoan phyla. Numbers of completely sequenced mitochondrial genomes are given. The shown phylogenetic relationships follow molecular studies by Schierwater et al. (2009). Modified symbols of Porifera, Cnidaria, Ctenophora, and Bilateria are courtesy of the Integration and Application Network (ian.umces.edu/symbols/). *From Osigus, H J., Eitel, M., Bernt, M., Donath, A., Schierwater, B., 2013. Mitogenomics at the base of Metazoa. Mol. Phylogenetics Evol. 69, 339–351.*

ventral sole to feed and digest the food, a feeding mode that is not observed in any other metazoan group except placozoans (Sperling et al., 2008; Sperling and Vinther, 2010). Keep in mind, however, that "there is still no agreement on the animal nature of *Dickinsonia* and related forms" (Buatois and Mángano, 2016).

References

Abedin, M., King, N., 2008. The premetazoan ancestry of cadherins. Science 319, 946–948.

Abedin, M., King, N., 2010. Diverse evolutionary paths to cell adhesion. Trends Cell Biol. 20, 734–742.

Adamidi, C., Wang, Y., Gruen, D., Mastrobuoni, G., You, X., Tolle, D., 2011. *De novo* assembly and validation of planaria transcriptome by massive parallel sequencing and shotgun proteomics. Genome Res. 21, 1193–1200.

Anderson, D.P., Whitney, D.S., Hanson-Smith, V., Woznica, A., Campodonico-Burnett, W., Volkman, B.F., et al., 2016. Evolution of an ancient protein function involved in organized multicellularity in animals. eLife 5 (2016), e10147.

Aouacheria, A., et al., 2006. Insights into early extracellular matrix evolution: spongin short chain collagen-related proteins are homologous to basement membrane type IV collagens and form a novel family widely distributed in invertebrates. Mol. Biol. Evol. 23, 2288–2302.

Banerjee, S., et al., 2006. Organization and function of septate junctions: an evolutionary perspective. Cell Biochem. Biophys. 46, 65–77.

Bing-Yu, M., et al., 1997. Morphological and functional studies on the epidermal cells of amphioxus (*Branchiostoma belcheri tsingtauense*) at different developmental stages. Chin. J. Oceanol. Limnol. 15, 236–241.

Boury-Esnault, N., et al., 2003. Larval development in the homoscleromorpha (Porifera, Demospongiae) invertebr. Biol. 122, 187–202.

Boute, N., et al., 1996. Type IV collagen in sponges, the missing link in basement membrane ubiquity. Biol. Cell. 88, 37–44.

Brower, D.L., et al., 1997. Molecular evolution of integrins: genes encoding integrin beta subunits from a coral and a sponge. Proc. Natl. Acad. Sci. U.S.A. 94, 9182–9187.

Brunet, T., Arendt, D., 2016. From damage response to action potentials: early evolution of neural and contractile modules in stem eukaryotes. Phil. Trans. R. Soc. B 371 (1685), 20150043.

Brunet, T., King, N., 2017. The origin of animal multicellularity and cell differentiation. Dev. Cell 43, 124–140.

Buatois, L.A., Mángano, M.G., 2016. Ediacaran ecosystems and the dawn of animals. In: Mángano, M.G., Buatois, L.A. (Eds.), The Trace-Fossil Record of Major Evolutionary Events, Topics in Geobiology 39. Springer, pp. 27–72. Chapter 2.

Butterfield, J., 2007. Macroevolution and macroecology through deep time. Palaeontology 50, 41–55.

Burkhardt, P., 2015. The origin and evolution of synaptic proteins – choanoflagellates lead the way. J. Experiment. Biol. 218, 506–514.

Cavalier-Smith, T., 2017. Origin of animal multicellularity: precursors, causes, consequences—the choanoflagellate/sponge transition, neurogenesis and the Cambrian explosion. Phil. Trans. R. Soc. B 372, 20150476.

Chapman, J.A., et al., 2010. The dynamic genome of *Hydra*. Nature 464, 592–596.

Clapham, D.E., 2007. Calcium signaling. Cell 131, 1047–1058.

Crick, F.H., 1968. The origin of the genetic code. J. Mol. Biol. 38, 367–369.

Dellaporta, S.L., Xu, A., Sagasser, S., Jakob, W., Moreno, M.A., et al., 2006. Mitochondrial genome of *Trichoplax adhaerens* supports placozoa as the basal lower metazoan phylum. Proc. Natl. Acad. Sci. U.S.A. 103, 8751–8756.

Eitel, M., Francis, W.R., Varoqueaux, F., Daraspe, J., Osigus, H.-J., Krebs, S., et al., 2018. Comparative genomics and the nature of placozoan species. PLoS Biol. 16 (9), e3000032.

Eitel, M., Guidi, L., Hadrys, H., Balsamo, M., Schierwater, B., 2011. New insights into placozoan sexual reproduction and development. PLoS One 6 (5), e19639.

Eme, L., Sharpe, S.C., Brown, M.W., Roger, A.J., 2014. On the age of eukaryotes: evaluating evidence from fossils and molecular clocks. Cold Spring Harb. Perspect. Biol. 6 (2014), a016139.

Ereskovsky, A.V., et al., 2009. The Homoscleromorph sponge *Oscarella lobularis*, a promising sponge model in evolutionary and developmental biology: model sponge *Oscarella lobularis*. Bioessays 31, 89–97.

Erwin, D.H., 2005. The origin of animal body plans. In: Briggs, D.E.G. (Ed.), Evolving Form and Function: Fossils and Development. Proceedings of a Symposium Honoring Adolf Seilacher for His Contributions to Paleontology, in Celebration of His 80th Birthday. Peabody Museum. April 1-2, 2005.

Fairclough, S.R., Chen, Z., Kramer, E., Zeng, Q., Young, S., Robertson, H.M., Begovic, E., et al., 2013. Premetazoan genome evolution and the regulation of cell differentiation in the choanoflagellate *Salpingoeca rosetta*. Genome Biol. 14 (2013), R15.

Fairclough, S.R., Dayel, M.J., King, N., 2010. Multicellular development in a choanoflagellate. Curr. Biol. 20, R875–R876.

Franzen, W., 1988. Oogenesis and larval development of *Scypha ciliata* (Porifera, Calcarea). Zoomorphology 107, 349—357.

Karpov, S.A., Coupe, S.J., 1998. A revision of choanoflagellate genera *Kentrosiga* Schiller, 1953 and *Desmarella* Kent, 1880. Acta Protozool. 37, 23—28.

King, N., Westbrook, M.J., Young, S.L., Kuo, A., Abedin, M., Chapman, J., et al., 2008. The genome of the choanoflagellate *Monosiga brevicollis* and the origin of metazoans. Nature 451, 783—788.

Kirk, D.L., 2005. A twelve-step program for evolving multicellularity and a division of labor. Bioessays 27, 299—310.

Ledger, P.W., 1975. Septate junctions in the calcareous sponge *Sycon ciliatum*. Tissue Cell 7, 13—18.

Leys, S.P., et al., 2009. Epithelia and integration in sponges. Integr. Comp. Biol. 49, 167—177.

Litvin, O., et al., 2006. What is hidden in the pannexin treasure trove: the sneak peek and the guesswork. J. Cell Mol. Med. 10, 613—634.

Love, A.C., Stewart, T., Wagner, G., Newman, S.A., 2017. Perspectives on integrating genetic and physical explanations of evolution and development: An introduction to the symposium. Integr. Comp. Biol. 57, 1258—1268.

Ludeman, D.A., Farrar, N., Riesgo, A., Paps, J., Leys, S.P., 2014. Evolutionary origins of sensation in metazoans: functional evidence for a new sensory organ in sponges. BMC Evol. Biol. 14, 3.

MacKie, G.O., Singla, C.L., 1983. Studies on Hexactinellid sponges. I. Histology of *rhabdocalyptus* Dawsoni (Lambe, 1873). Phil. Trans. Roy. Soc. Lond. B 301, 365—400.

Magie, C.R., Martindale, M.Q., 2008. Cell-cell adhesion in the cnidaria: insights into the evolution of tissue morphogenesis. Biol. Bull. 214, 218—232.

Maldonado, M., 2004. Choanoflagellates, choanocytes, and animal multicellularity. Invertebr. Biol. 123, 1—22.

Miller, D.J., Ball, E.E., 2005. Animal evolution: the enigmatic phylum revisited. Curr. Biol. 15, R26—R28.

Morris, S.C., 1998. Early metazoan evolution: reconciling paleontology and molecular biology. Am. Zool. 38, 867—877.

Muller, W.E., Muller, I.M., 2003. Analysis of the sponge [Porifera] gene repertoire: implications for the evolution of the metazoan body plan. Prog. Mol. Subcell. Biol. 37, 1—33.

Nakanishi, N., Stoupin, D., Degnan, S.M., Degnan, B.M., 2015. Sensory flask cells in sponge larvae regulate metamorphosis via calcium signaling. Integr. Comp. Biol. 55, 1018—1027.

Newman, S.A., Bhat, R., 2009. Dynamical patterning modules: a "pattern language" for development and evolution of multicellular form. Int. J. Dev. Biol. 53, 693—705.

Newman, S.A., Müller, G.B., 2001a. Epigenetic mechanisms of character origination. J. Exp. Zool. 288, 304—317.

Newman, S.A., Müller, G.B., 2001b. Morphological Evolution: Epigenetic Mechanisms. Encyclopedia Life Sc.

Nichols, S.A., et al., 2006. Early evolution of animal cell signaling and adhesion genes. Proc. Natl. Acad. Sci. U.S.A. 103, 12451—12456.

Nikitin, M., 2014. Bioinformatic prediction of *Trichoplax adhaerens* regulatory peptides. Gen. Comp. Endocrinol. 212, 145—155.

Niklas, K.J., Newman, S.A., 2013. The origins of multicellular organisms. Evol. Dev. 15, 41—52.

Osigus, H.-J., Eitel, M., Bernt, M., Donath, A., Schierwater, B., 2013. Mitogenomics at the base of Metazoa. Mol. Phylogenetics Evol. 69, 339—351.

Parfrey, L.W., Lahr, D.J., 2013. Multicellularity arose several times in the evolution of eukaryotes. Bioessays 35, 339—347.

Peterson, K.J., Butterfield, N.J., 2005. Origin of the Eumetazoa: testing ecological predictions of molecular clocks against the Proterozoic fossil record. Proc. Natl. Acad. Sci. U.S.A. 102, 9547–9552.

Philippe, H., Derelle, R., Lopez, P., Pick, K., Borchiellini, C., Boury-Esnault, N., et al., 2009. Phylogenomics revives traditional views on deep animal relationships. Curr. Biol. 19, 706–712.

Quastler, H., 1964. The Emergence of Biological Organization. Yale University, New Haven, CT, p. 16.

Sabella, C., et al., 2007. Cyclosporin A suspends transplantation reactions in the marine sponge *Microciona prolifera*. J. Immunol. 179, 5927–5935.

Sakarya, O., et al., 2007. A post-synaptic scaffold at the origin of the animal kingdom. PLoS One 1–9.

Schierwater, B., 2005. My favorite animal, *Trichoplax adhaerens*. Bioessays 27, 1294–1302.

Schierwater, B., Eitel, M., Jakob, W., Osigus, H.-J., Hadrys, H., Dellaporta, S.L., et al., 2009. Concatenated analysis sheds light on early metazoan evolution and fuels a modern "urmetazoon" hypothesis. PLoS Biol. 7 (1), e1000020.

Schierwater, B., Kamm, K., Srivastava, M., Rokhsar, D., Rosengarten, R.D., Dellaporta, S.L., 2008. The early ANTP gene repertoire: insights from the placozoan genome. PLoS One 3 (8), e2457.

Schleicherová, D., Dulias, K., Osigus, H.-J., Paknia, O., Hadrys, H., Schierwater, B., 2017. The most primitive metazoan animals, the placozoans, show high sensitivity to increasing ocean temperatures and acidities. Ecology and Evolution 7, 895–904.

Senatore, A., Raiss, H., Le, P., 2016. Physiology and evolution of voltage-gated calcium channels in early diverging animal phyla: Cnidaria, placozoa, Porifera and Ctenophora. Front. Physiol. 7 (2016), 481.

Senatore, A., Reese, T.S., Smith, C.L., 2017. Neuropeptidergic integration of behavior in *Trichoplax adhaerens*, an animal without synapses. J. Exp. Biol. 220, 3381–3390.

Shestopalov, V.I., Panchin, Y., 2008. Pannexins and gap junction protein diversity. Cell. Mol. Life Sci. 65, 376–394.

Signorovitch, A.Y., Dellaporta, S.L., Buss, L.W., 2005. Molecular signatures for sex in the Placozoa. Proc. Natl. Acad. Sci. U.S.A. 102, 15518–15522.

Smith, C.L., Abdallah, S., Wong, Y.Y., Le, P., Harracksingh, A.N., Artinian, L., et al., 2017. Evolutionary insights into T-type Ca^{2+} channel structure, function, and ion selectivity from the *Trichoplax adhaerens* homologue. J. Gen. Physiol. 149, 483–510.

Smith, C.L., Pivovarova, N., Reese, T.S., 2015. Coordinated feeding behavior in *Trichoplax*, an animal without synapses. PLoS One 10 (9), e0136098.

Smith, C.L., Varoqueaux, F., Kittelmann, M., Azzam, R.N., Cooper, B., Winters, C.A., et al., 2014. Novel cell types, neurosecretory cells and body plan of the early-diverging metazoan, *Trichoplax adhaerens*. Curr. Biol. 24, 1565–1572.

Sperling, E.A., Vinther, J., 2010. A placozoan affinity for Dickinsonia and the evolution of late Proterozoic metazoan feeding modes. Evol. Dev. 12, 201–209.

Sperling, E., Vinther, J., Pisani, D., Peterson, K., 2008. A placozoan affinity for Dickinsonia and the evolution of Late Precambrian metazoan feeding modes. In: Cusack, M., Owen, A., Clark, N. (Eds.), Programme with Abstracts, vol. 52. Palaeontological Association Annual Meeting., Glasgow, UK, p. 81.

Spiegel, I., Peles, E., 2002. Cellular junctions of myelinated nerves (Review). Mol. Membr. Biol. 19, 95–101.

Tyler, S., 2003. Epithelium-the primary building block for metazoan complexity. Integr. Comp. Biol. 43, 55–63.

Varoqueaux, F., Fasshauer, D., 2017. Getting nervous: an evolutionary overhaul for communication. Annu. Rev. Genet. 51, 455–476.

Voigt, O., Collins, A.G., Pearse, V.B., Pearse, J.S., Ender, A., Hadrys, H., Schierwater, B., 2004. Placozoa - no longer a phylum of one. Curr. Biol. 14, R944–R945.

Williams, E.A., Verasztó, C., Jasek, S., Conzelmann, M., Shahidi, R., Bauknecht, P., 2017. Synaptic and peptidergic connectome of a neurosecretory center in the annelid brain. eLife 6 (2017), e26349.

Woese, C.R., 1965. On the evolution of the genetic code. Proc. Natl. Acad. Sci. U.S.A. 54, 1546–1552.

Woollacott, R.M., Pinto, R.L., 1995. Flagellar basal apparatus and its utility in phylogenetic analyses of the Porifera. J. Morphol. 226, 247–265.

Further reading

Muller, G.B., 2007. Evo-devo: extending the evolutionary synthesis. Nat. Rev. Genet. 8, 943–950.

Chapter 2

Phanerozoic evolution—Ediacaran biota

Chapter outline

Ediacaran fauna—the prelude to the Cambrian explosion

Many researchers believe the metazoan life began during the Cryogenian era (720−635 Ma) or about 630 million years ago (Budd, 2008), followed by the Early Ediacaran (635−585 Ma) (Erwin, 2015). The Ediacaran macrofauna

Epigenetic Mechanisms of the Cambrian Explosion. https://doi.org/10.1016/B978-0-12-814311-7.00002-0
27

fossils appeared first in 579 Ma (Erwin, 2015), and most Ediacaran fossils are dated 575−541 Ma (Xiao and Laflamme, 2009).

Earlier students considered Ediacaran biota as a failed experiment in animal evolution, which represent self-organizing structures, inflated by mechanical forces rather than metazoans, or, at best, a new form of life, between plants and animals, hence proposed to classify them as a quite new group under the name Vendozoa/Vendobionta, unrelated to any extant group (Pflug, 1973; Seilacher, 1989, 1992; Cavalier-Smith, 2017). For others, they were cnidarian-grade animals (Conway Morris, 1993a,b), with some species/genera resembling extant organisms (e.g., *Kimberella* shows morphological similarities with extant mollusks). No sufficiently supported hypotheses about any possible relationship of Ediacaran fauna and extant animals have been possible to formulate, and the place of Ediacaran fauna in the tree of life basically remains unknown.

Based on molecular records, it is generally admitted that all pre-Cambrian lineages went extinct (Erwin et al., 2011) and no widely accepted phylogenetic relationships with extant taxa have been established (Fig. 2.1).

Fossil record

Among the best studied and widely recognized Ediacaran macrofossils are those of the genus *Dickinsonia*. Fossils of this genus look as flat, oval bilaterian organisms but lack any distinct organ. Based on the structure, physiology, and

Major Ediacaran fossil localities

Mackenzie Mountains,NW Canada Korbusunka River,Siberia, Russia Lantian biota, Auhui Province, China
Eastern Newfdld, Canada   Podolia, Ukraine   Columbia, Mato Grosso do Sul, Brazil
Chamwood Forest, UK   Weng'an Guizhou province, China Fann Aar, Namibia
White Sea, Russia   Yangtse Gorges region, china   Flinders Ranges, Ediacara and
                                Nilpena, Australia

FIGURE 2.1 Map of the world major fossil localities. *From Dunn, F.S., Liu, A.G., 2017. Fossil focus: the Ediacaran biota. Palaeontology Online 7 (1), 1−15. Internet. http://www. palaeontologyonline.com/articles/2017/fossil-focus-ediacaran-biota.*

behavior of the modern placozoa, it has been proposed that *Dickinsonia* may represent a placozoan-grade animal, but no confirmed or generally accepted placozoan fossils have been discovered to verify the hypothesis.

Dickinsonia grows by adding new segments or modules and by expansion of existing ones so that the body maintains its length—width proportions. *Dickinsonia* feeding traces indicate that it moved on Ediacaran matgrounds and used its ventral sole for feeding and digesting the food (Sperling et al., 2008; Sperling and Vinther, 2010). Locomotion excludes the possibility of algal or sponge origin of the *Dickinsonia* fossils. Moreover, among modern metazoans, external digestion is unique to placozoans, unknown in the rest of metazoan taxa.

Another hypothesis posits that *Dickinsonia* is ancestral to segmented bilaterians (Hoekzema et al., 2017). Two decades ago, Seilacher proposed that *Dickinsonia* belonged to an unknown kingdom of animals (Seilacher, 1992) that went extinct by the beginning of the Cambrian. Two other hypotheses posit plant and unicellular nature of *Dickinsonia*, but most recent research, especially the fact that *Dickinsonia* left feeding and body fossils (Figs. 2.2 and 2.3), rejects the algal nature and indicates that this iconic Ediacaran genus belongs to the Kingdom Animalia (Bobrovskiy et al., 2018). However, the phylogenetic relationships of the genus *Dickinsonia* remain still undetermined.

Besides sponges (Fig. 2.16), to Ediacaran fossils belong a dozen of other bilaterian phyla, but only genus *Kimberella* (probably a mollusk) has received broad acceptance (Peterson et al., 2009; Yin et al., 2015).

Kimberella fossils (Fig. 2.4) represent an Ediacaran bilaterian fossil genus comprising ∼ 100 ichnospecies. *Kimberella quadrata*, a bilaterian triploblastic

(A)   **(B)**

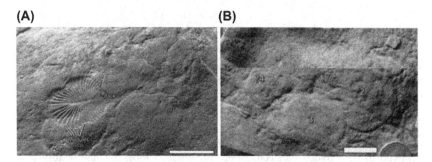

FIGURE 2.2 Body fossils of *Dickinsonia* and feeding traces. (A) Body fossil of *Dickinsonia costata* associated with a series of feeding traces. Numbers delineate the order of their formation in relation to the body fossil at the end of the series of traces (trace #3 made last). Note the distinct difference in relief between the trace fossils and the body fossil and the overlapping nature between the traces. Scale bar is 2 cm. (B) Circular series of traces preserving indications of modules and distinct overlap between the traces (previously figured by Gehling et al., 2005; SAM 40844). Along with previously figured specimens (Ivantsov and Malakhovskaya, 2002; Gehling et al., 2005; Fedonkin and Vickers-Rich, 2007) showing circular movements, this demonstrates that the tracks are not current-driven features. Scale bar is 2 cm. *From Sperling, E.A., Vinther, J., 2010. A placozoan affinity for Dickinsonia and the evolution of late Proterozoic metazoan feeding modes. Evol. Dev. 12, 201–209.*

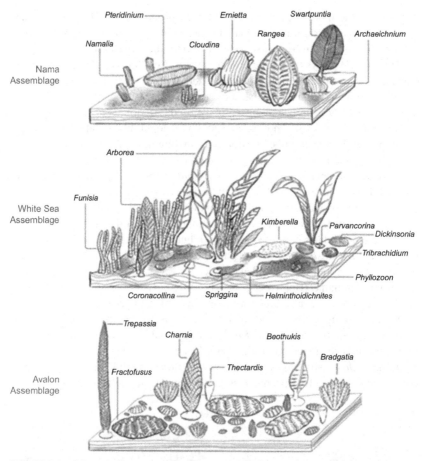

FIGURE 2.3 Schematized representation of community composition and relative abundance of taxa characteristic of the three assemblages of the Ediacaran biota. *From Droser, M.L., Tarhan, L.G., Gehling, J.G., 2017. The rise of animals in a changing environment: global ecological innovation in the late Ediacaran. Annu. Rev. Earth Planet Sci. 45, 593–617.*

trace fossil found more than a century ago in South Australia, was initially believed to be a hydrozoan (Glaesner and Wade, 1966), but it probably is a bilaterian animals (Butterfield, 2006).

The oldest of *Kimberella* body and trace fossils from the North Sea (Zimnie Gory) date 555–558 Ma and the organism was described as the oldest triploblastic trace fossil and "a bilaterian metazoan, more complex than a flatworm, more like a mollusk" (Martin et al., 2000; Fedonkin and Waggoner, 1997; Fedonkin et al., 2007), but the identification has been contended by other investigators (Sigwart and Sutton, 2007; Butterfield, 2006, 2008). However, whether *Kimberella* is a bilaterian or a coelenterate-grade animal is

FIGURE 2.4 *Kimberella* (*white arrow*) with *Radulichnus* grazing traces (*black arrow*). *From Xiao, S.H., Laflamme, M., 2009. On the eve of animal radiation: phylogeny, ecology and evolution of the Ediacara biota. Trends Ecol. Evol. 24 31—40.*

still unresolved (Budd and Jensen, 2017), so no firm relationship of *Kimberella* with any extant metazoan group has been possible to be established.

Spriggina is another bilaterian metazoan genus of the end of the Ediacaran, about 550 million years ago. No relationship has been possible to establish between *Spriggina* and any other extant or extinct animals (Fig. 2.5).

No consensus also exists among investigators on whether the tubular fossils assigned to *Sinocyclocyclicus, Quadratitubus, Ramitubus*, and *Crassitubus*, in Ediacaran Doushantuo Formation at Weng'an, South China, are cnidarian grade of closely related taxa (Liu et al., 2008), or represent fossilized algal cells (Cunningham et al., 2015; Liu et al., 2008), or cyanobacteria (Cunningham et al., 2015).

FIGURE 2.5 *Spriggina* with bilateral symmetry, anterior—posterior differentiation, and possible segmentation. *From Xiao, S.H., Laflamme, M., 2009. On the eve of animal radiation: phylogeny, ecology and evolution of the Ediacara biota. Trends Ecol. Evol. 24 31—40.*

A claim about an Ediacaran bilaterian animal preceding by ~40−50 million years the Cambrian explosion (Chen et al., 2004) has been contended because of failing to fully consider taphonomy (changes occurring after death) and diagenesis (changes in the sediment after deposition) (Bengtson and Budd, 2004) (Fig. 2.6).

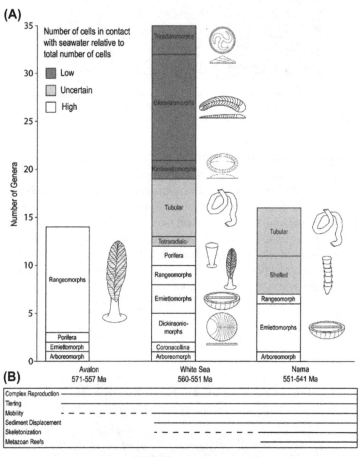

FIGURE 2.6 Morphological and ecological changes associated with the three fossil assemblages of the Ediacaran biota. (A) Stacked bar histogram of relative diversity through time of Ediacaran clades and color coded to reflect qualitative estimates of the relative number of cells in contact with seawater (see the text for detailed explanation of these assignments). *Line drawings* representing iconic fossil taxa of morphological groupings are shown when multiple taxa from that group are present in the fossil assemblage. Drawings not to scale. (B) Ecological innovations within the three assemblages. *Solid lines* are used when multiple taxa exhibit the specific behavior within a given assemblage. *Dotted lines* represent behaviors only known for a single taxa or trace fossil. Note the high overall diversity and number of taxa with low relative numbers of cells in direct contact with seawater and the appearance of several energetically demanding ecologies during the second wave (White Sea assemblage). *From Evans, S.D., Diamond, C.W., Droser, M.L., Lyons, T.W., 2018. Dynamic oxygen and coupled biological and ecological innovation during the second wave of the Ediacara Biota. Emerg. Topics Life Sci. 2, 223–233.*

The end of Ediacaran biota

Different hypotheses are presented about the causes of the disappearance of the Ediacaran biota around the Ediacaran—Cambrian border. The most popular among these is the "biotic replacement" hypothesis, which posits that the soft-body Ediacaran animals disappeared gradually as a result of competition by emerging Cambrian fauna (Darroch et al., 2015, 2016) or predation by Cambrian predators (Laflamme et al., 2013) with complex sense organs and information processing systems and effector organs for both capturing prey and avoiding predation (Wray, 2015). Another hypothesis considers the disappearance of Ediacaran fauna as a result of a geochemical catastrophe, an abrupt environmental change to which Ediacaran biota could not adapt. Other biologists believe that the above hypotheses do not necessarily exclude each other, and competition-determined replacement stage preceded the geochemically determined extinction stage (Smith et al., 2016). According to a third hypothesis, Ediacaran biota was eliminated as a result of the loss of microbial matgrounds by the end of the Ediacaran eon. A more recent hypothesis posits that the brief period of oxygenation of the ocean during Ediacaran period (579—550 Ma) was followed by anoxia of the ocean at the end of the Ediacaran (Zhang et al., 2018).

Sponges (Porifera)

Sponges (Porifera) are another primitive phylum of extant metazoans derived from pre-Cambrian stem lineages. The phylum comprises 5000—10,000 aquatic species, attached permanently to solid substrates. Sponges are exceptionally diverse as far as the morphology and size is concerned. They have no body symmetry, tissues, or organs. They lack muscle cells and neurons, but have 12 cell types (choanocytes, amoebocytes, pinacocytes, porocytes, sclerocytes, myocytes, oocytes, sperm cells, lophocytes, collencytes, and rhabdiferous cells).

Fossil record

Sponges belonging to the earliest pre-Cambrian animals are indicated not only by their extremely simple structure, consisting of only two cell layers, but also by the pre-Cambrian (early Vendian) presence of sponges and their larvae in the fossil record in South China, dated about 580 Ma (Nichols et al., 2006). The latest paleontological record also pushed back the time of emergence of sponges to 600 Ma, based on the discovery of a 3D-preserved 1 millimeter across sponge fossil, *Eocyathispongia qiania*, in Doushantuo strata, China (Yin et al., 2015) (Fig. 2.7). Like extant Porifera, fossilized sponges are asymmetrical and have no body axes, and the walls have openings that form a canal system through which the water flows in.

(A) (B)

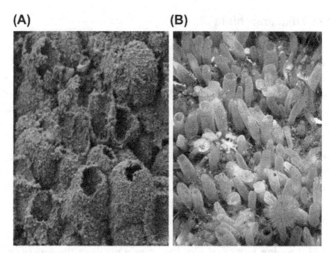

FIGURE 2.7 Specialized surface structures. (A) Scanning electron micrograph of the fossil exterior with the framed area showing the location of a patch of short, hollow tubes. (B) Surface of the extant demosponge *Polymastia penicillus* showing hollow papillae up to 1 cm long. *From Yin, Z., Zhu, M., Davidson, E.H., Bottjer, D.J., Zhao, F., Tafforeau, P., 2015. Sponge grade body fossil with cellular resolution dating 60 Myr before the Cambrian. Proc. Natl. Acad. Sci. U.S.A 112, E1453–60.*

Based on the discovery of sedimentary biomarkers, cholestanes, and hydrocarbon remains of C_{30} sterols as well as the suggestion that steroid 26-methylstigmastane (Zumberge et al., 2018) was produced by marine demosponges in Oman, in strata dated >635 Ma, it was concluded that the first demosponges appeared about 635 Ma (Love et al., 2009; Gold et al., 2016). Moreover, it was proposed that demosponges, and hence multicellular animals, were not rare, at least since the Cryogenian period, ca. 750 Ma. Another investigator dated the fossils to 645 Ma but identified them of algal origin (Antcliffe, 2013). Other biologists' record pushed the sponge appearance deeper in the Cryogenian period ($\sim$750 Ma). These biomarkers preserved in rocks were thought to be derived via degradation of steroids and sterols, hence were regarded as molecular evidence for eukaryotic cells (Xiao, 2013). And finally, there is unconfirmed evidence that microfossils, termed *Otavia antiqua* gen. et sp. nov., evolved before the Minoan glaciation, about 760 Ma (Brain et al., 2012; Zumberge et al., 2018). It is also reported that a sponge grade body fossil dated to 60 Ma before the Cambrian was discovered in 2015 (Yin et al., 2015).

However, the most recent study seems to have invalidated the hypothesis that sponges evolved more than 100 Ma before the Cambrian because the so-called sponge biomarkers, C30-steranes, 26-methylstigmastane (26-mes), 24-isopropylcholestane (24-ipc), 24-n-propylcholestane (24-npc) 26-mes, in fossil record are a distinctive character of sponges. The study demonstrated that such biomarkers were also produced by *Protista* spp. *Rhizaria* and possibly other groups of animals and plants that populated seas earlier than sponges (Nettersheim et al., 2019) (Fig. 2.8).

In all likelihood, the most reliable biomarker of fossil sponges is the presence of spicules, which lacks in Ediacaran deposits, despite the abundant amounts of silica in sea that would favor spicule preservation (Nettersheim et al., 2019). This is a clearly more reliable criterion that will bring the appearance of sponges closer to the Cambrian, about 560 Ma. Mongolian silica hexactinellid sponges from ~545 Ma, at the pre-Cambrian—Cambrian boundary, are the oldest widely accepted sponge fossil, whereas the oldest reliable sponge fossil is dated 535 Ma. This species with siliceous spicules from the basal Cambrian is found in Soltanieh Formation, Iran (Antcliffe et al., 2014).

Morphology

Sponges are extraordinarily diverse organisms. They have no body symmetry, tissues, or organs. They lack muscle cells and neurons, but have at least 12 other cell types: archaeocytes, amoebocytes, choanocytes, pinacocytes, sclerocytes, myocytes (compared to the smooth muscle cells), collencytes, lophocytes, porocytes, oocytes, and sperm cells, or even 16 cell types (without mentioning subtypes) (Leys, 2015).

Based on the general morphology, sponges are divided in asconoid sponges looking as tubes with an opening to the outside, with walls perforated by pores ending in the internal cavity or spongocoel; *syconoids* that resemble asconoids but are larger and have thicker walls; *leuconoids*, which are more voluminous and are perforated by numerous canals ending to small chambers lined with choanocytes, which, by beating their flagella, force the water into the spongocoel and out of the osculum (Fig. 2.9).

Sponges respond with high morphological plasticity to variation in environmental conditions and the plasticity may be related to the low degree of complexity of the control system (see later control systems in eumetazoans). Sponges evolved before the Cambrian explosion, but their morphology remained little changed in the course of their long phylogeny (Nichols et al., 2006).

Reproduction

Sponges reproduce both asexually and sexually. Asexual reproduction comprises budding, fission, and formation of gemmules.

Most sponges are hermaphroditic and during the sexual reproduction the sponge forms both sperm cells and ova. Sperm cells are released in the water and upon encountering another sponge they enter the body and are transported by archaeocytes to mesoglea where they fertilize the eggs forming zygote, which develops into a ciliated larva that is released into the water and, after settling on a solid substrate, develops into a juvenile larva and the mature sponge.

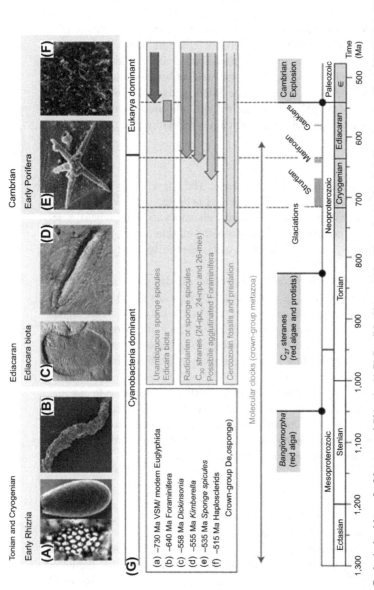

FIGURE 2.8 Geological evidence for Metazoa and *Rhizaria*. (A,B) Possible *Rhizaria*. Vase-shaped microfossil (left) and modern Euglyphid (right). *Rhizaria* can be traced to pre-Cryogenian and Cryogenian rocks, consistent with the Cryogenian appearance of C_{30} steranes. Images adapted from Porter et al., 2003. (C,D) Macroscopic fossils of the Ediacaran biota reflect the late Ediacaran emergence of animals Bobrovskiy et al., 2018. (E,F) Early Porifera (sponges). A sponge source for C_{30} steranes would require ubiquitous demosponge communities since the Cryogenian period, which is incompatible with the most recent interpretation of the sponge fossil record Botting and Muir, 2017. (E) Image adapted from Chang et al., 2017. (F) Image adapted from Botting et al., 2015. (G) Molecular clocks provide widely divergent estimates for the appearance of crown Metazoa (dos Reis et al., 2015), which range from 55 to 740 Myr before the oldest diagnostic metazoan fossils (Erwin et al., 2011; Peterson et al., 2004; Bobrovskiy et al., 2018). Є, Cambrian period. See main text and supplementary information for further details. *Figure credit: S. Porter (A, left); R. Meisterfeld (A, right); S. Pruss (B); S. Chang (E); J. Botting (F). From Nettersheim, B.J., Brocks, J.J., Schwelm, A., Hope, J.M., Not, F., Lomas, M., et al., 2019. Putative sponge biomarkers in unicellular Rhizaria question an early rise of animals Nature. Ecol. Evol. 3, 577–581.*

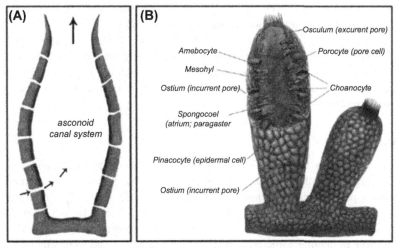

FIGURE 2.9   Modern sponge anatomy. (A) Schematic cross section of simple asconoid sponge morphology with a central cavity and base. Pores (ostia) in body wall carry water inflow, and the large orifice (osculum) is used for outflow (*arrows* indicate direction of water flow). (B) Schematic cross section showing three layers of the body wall, including external pinacocyte cells, internal choanocyte cells, and mesohyl separating them; incurrent pores in the body wall (ostia); and an excurrent opening (osculum) to the body chamber (spongocoel), as well as additional cell types (porocyte and amebocyte). *From Zongjun, Y., Maoyan, Z., Davidson, E.H., Bottjer, D., Zhao, F., Tafforeau, P., (2015). Sponge-grade body fossil with cellular resolution dating 60 Myr before the Cambrian. Proc. Natl Acad. Sci. 112, E1453–60.*

Evolution

The fact that sponges belong to the earliest pre-Cambrian animals is proven not only by their extremely simple structure, consisting of only two cell layers, but also by the discovery of the pre-Cambrian (early Vendian) existence of sponge fossils and their larvae about 580 mya, in South China (Li et al., 1998).

Based on results of their investigation of "signaling and adhesion genes that interact to coordinate complex morphogenetic events in eumetazoans as diverse as flies, worms, and humans," Nichols et al. (2006) conclude that "developmentally important signaling and adhesion gene families evolved before the divergence of sponge and eumetazoan lineages and were, therefore, in place in all animal lineages at the onset of the Cambrian explosion. Sponges entered the historic period of morphological radiation equipped with the adhesion machinery used by eumetazoans to selectively sort differentiated cell populations and to form epithelial barriers between differentiating body compartments" (Nichols et al., 2006).

The fact that sponges failed to evolve into more complex structures and remained a "dead end" of animal evolution indicates that the mere presence of the signaling genes, of the signal transduction pathways, and of the genetic toolkit, while being necessary conditions, is not sufficient for progressive evolution and increased structural and functional complexity.

If Porifera evolved before eumetazoans with nervous system, as most biologists believe, the question arises: why they entered this 600 million year-long evolutionary stasis to reach our time almost unchanged and stuck in an evolutionary dead end? Could the lack of a nervous system be the cause of their evolutionary dead end?

Development

According to the hypothesis of *generic factors*, the development and body patterning of simplest multicellular animals, such as sponges, may be influenced by physical mechanisms, such as differential adhesivity of different cell types, biochemical diffusion, viscoelasticity, etc. These factors may be responsible for formation of the two cell layers in sponges as well as for the formation of tubular structures, lumina, rods, spheres, and generally three-dimensional patterning (Newman, and Müeller, 2001).

The importance of environmental factors in sponge development is reflected in the absence of a strictly determined morphology, prevalence of irregular forms, and in the absence of clear limits of growth, compared to eumetazoans. This also is the reason for the wide use of other elements, such as cell features, reproduction modes, patterns of development, skeletal form, geometry of spicules, etc., in determining taxonomic relationships in sponges.

It is generally accepted the hypothesis of gastrula as a developmental stage of sponges was propounded more than a century ago by Ernst Haeckel (1834–1919). After him, other biologists made some progress in demonstrating the reality of the gastrula stage in sponges by showing ingression reorganization of tissues during embryogenesis. Now we know that after six cleavages, a 64-cell amphiblastula forms with a central cavity and ciliated cells in the anterior hemisphere, which evert to become the inner cell layer before the metamorphosis takes place (Leys and Eerkes-Medrano, 2005). However, for other authors, the gastrula stage occurs later, with the development of an internal cell layer by epithelial-mesenchyme transitions and transformation of larval ciliated cells (Nakanishi et al., 2014).

A number of eumetazoan attributes make it tempting to infer homology of gastrulation across metazoans, and the development of sponges and cnidarian suggests that gastrulation evolved through cell migration rather than invagination (Wörheide et al., 2012) (Fig. 2.10).

In sponges are not observed developmental signal cascades like those described in eumetazoans and there is little information on the control system in sponges. The high plasticity results from environmental factors (Meroz-Fine et al., 2005). However, theoretically at least, sponges cannot be devoid of an intrinsic pre- or nonneural control system. Neuroid phenomena observed in sponges point to the existence of such a control system for the development

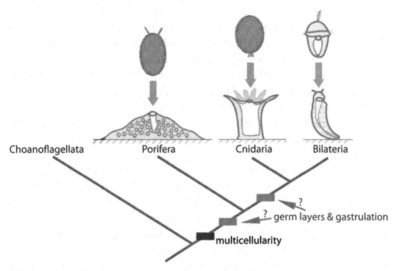

FIGURE 2.10 Evolution of germ layers and gastrulation. Animal multicellularity, with the larval and juvenile/adult body plans of extant metazoans possessing multiple cell layers, is depicted above. The *arrow* represents metamorphosis from larval to juvenile/adult forms. Cell layers are colored: orange (gray in print version), inner layer; green (dark gray in print version), outer layer; and blue (white in print version), middle layer. In the case of cnidarians and bilaterians, these correspond to endoderm, ectoderm, and mesoderm (bilaterians only), respectively. The phylogenetic relationship of these clades and the sister group to metazoans, the choanoflagellates, is shown below, along with the evolutionary origin of multicellularity. The origin of germ layers and gastrulation is debatable, occurring either before or after the divergence of sponges and eumetazoans. *From Nakanishi, N., Sogabe, S., Degnan, B.M., 2014. Evolutionary origin of gastrulation: insights from sponge development. BMC Biol. 12 (2014), 26.*

and maintenance of sponge structure and morphology. Indeed, studies on *Oscarella carmela* have shown that this sponge expresses many of the signaling genes and has all but one (nuclear hormone receptor pathway) of the seven major bilaterian signaling pathways: Wnt, TGF-p, Hedgehog, receptor tyrosine kinase, Jak/STAT, and Notch signaling pathways. This indicates that the mechanism of signal transduction pathways evolved before the evolution of eumetazoans, such as cnidarians and ctenophores, and before the Cambrian explosion. In bilaterians, these pathways are responsible for limb development, eye development, vertebrate segmentation, and assembly of neural circuits (Nichols et al., 2006). The evidence on the presence of miRNAs in sponges is met with skepticism (Thomson and Dinger, 2016).

The control system and behavior in sponges

Although sponges lack neurons and the neuroendocrine system, which is in the center of the integrated control system in eumetazoans, they have a control system that may memorize and direct the development.

The sponge *Amphimedon queenslandica* expresses both Delta and its receptor Notch gene, as well as a sponge bHLH gene, which shows strong proneural activity, and is involved in molecular mechanisms similar to those observed in bilaterian primary neurogenesis (Richards et al., 2008). Larvae of *Reniera* sponges respond rapidly to the increase in the intensity of light in their environment and to other external stimuli (Leys and Degnan, 2001).

In sponges are observed impulse conduction phenomena. Local tactile and electrical stimulation arrests exhalant water current in the sponge, *Rhabdocalyptus dawsoni*, by inducing an action potential that propagates throughout their body with a threshold of excitability of up to 30 s and a propagation velocity of 0.22 cm/s. The medium through which the action potential propagates is a syncytium, a layer of cytoplasmic homogenous cytoplasm (Leys et al., 1999). In some sponges, the spread of signals along the syncytium involves calcium-dependent communication between cells (Jacobs et al., 2007). According to one hypothesis cilia are the sensory and the osculum is their sensory organ that coordinates systemic sponge responses (Ludeman et al., 2014).

Despite lacking neurons and nerve net, larvae of the demosponge *A. queenslandica* show phototaxis by moving in the direction of light by rotating around the anterior—posterior axis. Changes in the intensity of light are sensed by a photoreceptive pigment contained in the cell cilia (Leys et al., 2002). The photopigments are identified as cryptochromes AqCry1 and AqCry2, of which the latter functions as a neuronless, opsinless eye in sponges (Rivera et al., 2012). Each cell may respond individually to light change or a rapid coordination may result from communication via cytoplasmic bridges among these cells (Leys and Bernard, 2005).

Flask cells of *A. queenslandica* are epithelial sensory secretory cells with a cilium (Sakarya et al., 2007). These sensory cells are found in the anterior third of the larval body. They express five of the postsynaptic proteins. They have a receptor and a signaling system that recognizes and responds to environmental stimuli by converting environmental stimuli into internal signals for inducing metamorphosis. The mechanism of conversion involves elevation of the intracellular Ca^{2+} levels in flask cells, leading to a cascade of changes in cellular behavior and state that determines larval settlement and metamorphosis (Nakanishi et al., 2015). Experimental use of a cnidarian GLWa-mide neuropeptides stimulates larval settlement in two coral reef sponges, *Cosocinoderma matthewsi* and *Rhopaloeides odorabile*, demonstrating chemosensory and potentially neural capabilities of these sponge species (Whalan et al., 2012). Sponges express almost all the postsynaptic scaffolding, especially in the flask cells of larval demosponges. In the sponge *A. queenslandica*, enzymes necessary for the processing and secretion of proneuropeptides are detected (Srivastava et al., 2010).

In sponges are observed many molecular signatures and signaling pathways of the neurons and the nervous systems, such as "adrenergic, adenosynergic, and glycinergic pathways, as well as pathways based on NO and extracellular cAMP are candidates for the regulation and timing of the endogenous contraction rhythm within pacemaker cells, while GABA, glutamate, and serotonin are candidates for the direct coordination of the contractile cells" (Ellwanger and Nickel, 2006), which may be central to the function of the control system in metazoans.

Behavioral repertoire of sponges is surprisingly rich considering the fact that they lack a nervous system. The freshwater sponge *Ephydatia muelleri* responds to mechanical stimuli with a series of peristaltic contractions that discharge water and wastes to the external environment (Elliott and Leys, 2007), whereas as larva the sponge responds to photic stimuli by changing the direction of swimming in opposing direction of the light (Leys and Degnan, 2001). It is believed that the sensory cells are responsible for sponge's responses.

Evolution of the neuron: the second informational revolution

Communication with hormones and other signaling molecules in sponges (Leys and Meech, 2011) and with neuropeptides in placozoans (Nikitin, 2015) evolved early in these organisms, while communication of cells via the gap junctions is strictly local (Feldman, 2010). But these forms of communication were too slow and local to meet the requirements of both the preys and predators that appeared at the Ediacaran—Cambrian border. This stimulated a selective pressure for better and faster communication between the neighboring cells and also with remote cells throughout the metazoan body.

This selective pressure led to the evolution of neurons in eumetazoans, cnidarians, and bilaterians. Neurons are cells specialized in receiving and transmitting information about the internal and external environment via synapses to other neurons and to effector cells/organs. The advent of the neuron was crucial for evolution of metazoan behavioral, physiological, and morphological diversification. Neurons are the basic units of the nervous system, and no eumetazoan, as we know them, would evolve and exist without the nervous system.

The evolution and differentiation of a metazoan cell into a neuron was a result of the integration into a single cell of voltage-gated elements, synaptic molecular machinery, and branching structures of dendrites and axons, to connect and exchange information with other neurons or other cells. The evolution of the neuron was a process rather than an event. In all likelihood, this occurred during the Ediacaran eon, with the evolution of the cnidarians or cnidarian-grade organisms.

There is reason to think that synapses may have evolved before axons and dendrites because the small size of the early organisms allowed axonless communication between closely associated neurons and both axons and dendrites use synapses for intercommunication. The view that the evolution of neurons started with evolution of synapses is known as "synapse-first hypothesis" (Ryan and Grant, 2009).

Evolution of the synapse

Synapses are fundamental and indispensable structures for the evolution of the neurons and the nervous system that is argued to have evolved before the evolution of dendrites and axons. Many mammalian synaptic protein components are found in unicellular eukaryotes and in nerveless metazoans (sponges and placozoa). The discovery of neuronally relevant components in unicellulars suggests that evolution of the neuron was not a fortuitous evolutionary event and it likely evolved more than once (Kristan, 2016).

Synaptic protein building blocks appeared first in prokaryote unicellulars as parts of mechanisms for sensing and responding to environmental stimuli, from which simple protosynapses, the precursors of complex neuronal synapses, evolved (Emes and Grant, 2012) (Fig. 2.11).

Synaptic proteins in unicellulars

Of all unicellulars, the most closely related to metazoans seem to be choanoflagellates (*Choanoflagellatea*) (King et al., 2008) and protists living as unicellular and colonial forms in marine and freshwater. Cells resembling choanoflagellates are found in sponges (choanocytes) and other multicellular animals.

Among the synaptic proteins identified in the unicellular choanoflagellate, *Monosiga brevicollis*, which are absent in metazoans, are a number of tyrosine kinases, which in this protist may play a role in detecting and adaptively responding to changes in environment, as well as adhesion protein homologs, such as cadherins, which may be used for prey capture (King et al., 2003). In choanoflagellates are identified neurally important synaptic proteins that are absent in nonmetazoans. However, the most ancient synaptic protein families are conserved in unicellular eukaryotes, such as the yeast *Saccharomyces cerevisiae* and the ameba *Dictyostelium discoideum*. Over 21% of the mammalian MASC (MAGUK-associated signaling complexes) genes and 25% of the postsynaptic density genes have orthologs in these protosynaptic nonmetazoan organisms (Ryan and Grant, 2009). A general view of the evolution of postsynaptic components in the living world is presented in Fig. 2.12.

Synaptic proteins in nerveless metazoans

Sponges have no neurons, but in the sponge *O. carmela*, besides tyrosine kinases are identified ankyrin and neurexin, components of eumetazoan synaptogenesis (Ryan and Grant, 2009). Choanoflagellates *M. brevicollis* and *Salpingoeca rosetta* have protosynaptic protein homologs of eumetazoan synapses, including neuronal scaffold proteins Homer, Shank, and DLG (Discs-large) (Burkhardt and Sprecher, 2017; Liebeskind et al., 2017), which are essential for fast transmission of messages from the cell membrane to the nucleus. Nevertheless, it seems that in these unicellulars, "neuronal" proteins

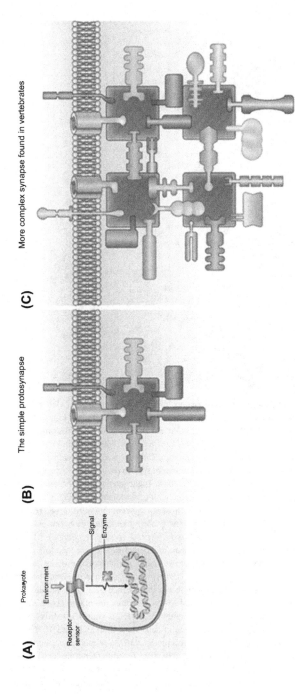

FIGURE 2.11 Sensing of the environment by prokaryotes and vertebrate synapses. In the prokaryote (A) the external environment (blue [white in print version]) may be a nutrient or diffusible signal that stimulates membrane proteins (receptor sensor) that can trigger intracellular kinases (enzyme) that regulate transcription. Multiple varieties of receptors and intracellular enzymes produce synaptic diversity. The basic components of the core signaling complexes found in eukaryotes (B) have been multiplied by the process of gene duplication, producing greater varieties of component proteins for vertebrate synapses (C). *From Emes, R.D., Grant, S.G.N., 2012. Evolution of synapse complexity and diversity. Annu. Rev. Neurosci. 35, 111–131 (combined figures).*

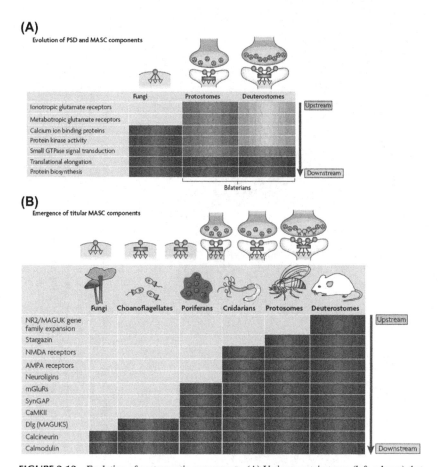

(A)
Evolution of PSD and MASC components

	Fungi	Protostomes	Deuterostomes
Ionotropic glutamate receptors			
Metabotropic glutamate receptors			
Calcium ion binding proteins			
Protein kinase activity			
Small GTPase signal transduction			
Translational elongation			
Protein biosynthesis			

Upstream

Downstream

Bilaterians

(B)
Emergence of titular MASC components

	Fungi	Choanoflagellates	Poriferans	Cnidarians	Protostomes	Deuterostomes
NR2/MAGUK gene family expansion						
Stargazin						
NMDA receptors						
AMPA receptors						
Neuroligins						
mGluRs						
SynGAP						
CaMKII						
Dlg (MAGUKS)						
Calcineurin						
Calmodulin						

Upstream

Downstream

FIGURE 2.12 Evolution of postsynaptic components. (A) Various protein types (left column) that constitute the postsynaptic density (PSD) and membrane-associated guanylate kinase (MAGUK) associated signaling complexes (MASCs) in unicellular eukaryotes (fungi), protostomes, and deuterostomes are ordered based on whether they have "upstream" or "downstream" signaling roles. Noncolored fields represent the absence of a given protein. *Dark gray rectangles* represent presence of protein. *Gray rectangles represent* enrichment of a protein type in protostomes or protostomes and deuterostomes when compared with unicellular eukaryotes. Light gray *rectangles* represent enrichment of a protein type in deuterostomes when compared with protostomes. The diagrams above each column represent the molecular assembly of MASC, in which upstream proteins (*blue [gray in print version] circles*) are connected to downstream proteins (*yellow [white in print version] triangles*) through intermediate signaling proteins (*red [dark gray in print version] rectangle*). The relative proportions of these proteins in eukaryotes, protostomes, and deuterostomes are therefore illustrated. (B) The emergence of titular MASC components across clades is illustrated. Proteins are ordered based on whether they are located "upstream" or "downstream" in synaptic signal transduction pathways (Emes et al., 2008). Noncolored fields represent the absence of a given protein, whereas *dark gray rectangles* denote its presence. Diagrams of MASC structure are placed above each clade, along with an illustration of a representative model organism. *AMPA*, α-amino-3-hydroxy-5-methyl-4-isoxazolepropionic acid; *CaMKII*, calcium/calmodulin-dependent protein kinase II; *Dlg*, discs-large homolog; *mGluRs*, metabotropic glutamate receptors; *NMDA*, N-methyl-D-aspartate; *NR2*, NMDA receptor 2; *SynGAP*, synaptic Ras GTPase activating protein. *Diagrams in part (A) are modified, with permission, from REF. 16© (2008) Macmillan Publishers Ltd. All rights reserved. From Ryan, T.J., Grant, S.G.N., 2009. The origin and evolution of synapses. Nat. Rev. Neurosci. 10, 701–712.*

perform nonsynaptic functions, whereas protosynaptic proteins emerged and their number expanded in the course of evolution (Burkhardt and Sprecher, 2017). The fact that the synaptic machinery of neurons existed before the evolution of cnidaria and bilateria is suggested by the evidence that neuron might have independently evolved 10–12 times in metazoans (Moroz, 2012).

Interesting in relation to the origin of the neuron is the discovery of the role of the *Trichoplax* gland cells as secretory cells that "could control locomotor and feeding behavior" of the animal (Smith et al., 2014). These neuron-like properties should be considered along the fact that *Trichoplax adhaerens* encodes for proteins, which are very homologous to proteins that participate in regulation of secretion in Neuralia, such as SNARE (soluble NSF attachment protein receptors) proteins, syntaxin-1, synaptobrevin, SNAP-25, and synapsin, a membrane protein of synaptic vesicles (Smith et al., 2014) and a neuronal protein that binds SNARE proteins.

On the other side of the synaptic gap (cleft) is the presynaptic density composed of a complex network of proteins, where electrical impulses stimulate the Ca^{2+}-sensitive machinery of exocytosis to secrete neurotransmitters (Senatore et al., 2016) that stimulate postsynaptic machinery.

Evolution of the action potential and voltage-gated ionic channels

The right concentration and proportions and the affinity of the interactions of postsynaptic proteins are necessary for the synapse assembly and may drive self-assembly (Conaco et al., 2012). It is thought that voltage-gated K^+ were first to evolve, but K_v channel alone cannot generate action potentials (Kristan, 2016). The second to evolve was Ca_v. Both Ca_v and Na_v channels evolved by a double duplication of the K_v gene, and by acquiring Ca_v and Na_v channels in proper combinations, metazoan cells became able to produce action potentials (Kristan, 2016).

Voltage-gated K^+ (Kv) channel is a tetramer membrane protein, whose pore is gated (opened) when the membrane potential (difference between the interior and exterior of the cell) increases to a particular threshold, which causes gates to open leading to an influx of Na^+, which in turn increases further the membrane potential and opens more gates for other ions, ultimately leading to the reverse process of closing the gates, arrest of Na^+ influx, opening of K^+ gates, and exit of K^+. These rises and falls of the membrane potential represent action potentials that are generated and propagated by neurons in response to various stimuli.

Cells are not a good medium of conduction of electricity, but neurons evolved a mechanism of transmitting electric information from the presynaptic neuron to the postsynaptic neuron via action potentials or spikes. The information is provided in the form of the number, temporal patterns, and duration of spikes. Transferred information in the postsynaptic neuron is processed and converted into chemical information and then reconverted back to electronic information (Arabzadeh et al., 2006; Stafford, 2010). Action potentials occur when various stimuli cause a difference between the electric potential inside and outside the cell wall. As a consequence of the opening of the voltage-gated ion channels, the cell membrane lets Ca^{2+} and Na^+ ions enter the cell.

From the beginning, eukaryote cells evolved mechanisms for maintaining their physiological chemistry against the excessive influx of Ca^{2+} and Na^+ ions by respective cation pumps (Brunet and Arendt, 2016). It is suggested that neurons may operate as hybrid digital—analog processor that along the firing/not firing action potentials process information via a range of voltages independently of spikes.

Besides neurons, action potentials occur in muscle cells and some other cell types, but they also occur in animal and plant unicellular eukaryotes. The fact that action potentials are recorded in various neuronal and myocyte types led biologists to the hypothesis that both neurons and myocytes differentiated from a common precursor mechanosensory-contractile myoepithelial cell with sensory, secretory, and contractile functions via division of labor, with the neurons retaining the depolarization—secretory function and myocytes the depolarization—contraction function. Myocytes convert calcium signals into contractions (Brunet and Arendt, 2016).

It is hypothesized that the signaling role of Ca^{2+}-induced depolarization in neuron and muscle cells or depolarization—contraction—secretion evolved early in the course of evolution as a response to the detrimental rise in the Ca^{2+} level, when the cell membrane was damaged (Brunet and Arendt, 2016). Accordingly, drastic increases in the Ca^{2+} concentration within the eukaryote cell caused membrane damage, to which the cell responded by contraction of the actomyosin ring around the damaged site and release of vesicles with sealing material (Fig. 2.13), in analogy with the release of neurotransmitters by

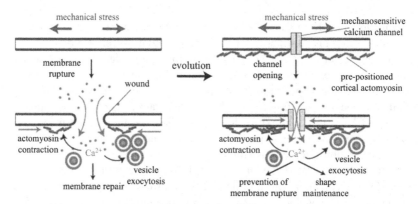

FIGURE 2.13 Emergence of mechanosensitive Ca^{2+} channels and cortical actomyosin for anticipating membrane damage in stem eukaryotes. *From Brunet, T., Arendt, D., 2016. From damage response to action potentials: early evolution of neuronal and contractile modules in stem eukaryotes. Philos. Trans. R. Soc. Lond. Biol. Sci. 371, 20150043.*

the presynaptic side of the synapse. This process of the membrane repair involved both contraction of actomyosin ring and secretion of the sealing material. Owing to its crucial role in the physiology and the integrity of the eukaryote cell, the repair mechanism evolved early in the LECA (last eukaryotic common ancestor). Voltage-gated sodium (Na_V) and calcium (Ca_V) channels are closed when the membrane potential (the difference between the electric potentials inside and outside the cell membrane) is at or below the so-called resting potential. Any increase of the membrane potential above a threshold causes gates to open, leading to an influx of the Ca^{2+} and Na^+ in the cytoplasm. Increase of the transmembrane potential causes the channels to open, thus allowing cations to enter the cell.

The demosponge *A. queenslandica* has almost the full set of synaptic genes and even organized into an independent community, suggesting that synapses evolved via exaptation of existing genes and smaller postsynaptic protein communities (Conaco et al., 2012).

On the path to differentiation of the neuron

Sponges have no recognizable neurons and nervous system, but the demo-sponge *A. queenslandica* larvae have epithelial sensory secretory flask cells with a cilium (Sakarya et al., 2007), which may have served as precursors of the evolution of neurons (Sakarya et al., 2007). These sensory cells are found in the anterior third of the larval body. They have a receptor and a signaling system that recognizes and responds to environmental stimuli by converting environmental stimuli into internal signals for inducing metamorphosis. The mechanism of conversion involves elevation of the intracellular Ca^{2+} levels, leading to a cascade of changes in cellular behavior and state that determines larval settlement and metamorphosis (Nakanishi et al., 2015), which remind us of the role of the nervous system in controlling metamorphosis in eumeta-zoans. Moreover, as mentioned earlier, experimental use of a cnidarian GLWamide neuropeptide stimulates larval settlement in two coral reef sponges, *C. matthewsi* and *R. odorabile*, demonstrating chemosensory and potentially neural capabilities of these sponge species (Whalan et al., 2012).

It is worthwhile mentioning that besides synaptic proteins, in neuronless metazoans like placozoans are found neuropeptides, suggesting that they were present in the common ancestor of placozoans and eumetazoans and predate the advent of the nervous system. Neuropeptide-like proteins in *T. adhaerens* may be involved in paracrine regulation of behaviors like ciliary crawling of the animal and physiological processes like secretion of digestive enzymes (Jékely, 2013). It is interesting to note that even typical neuronal products like neurotransmitters are discovered in unicellular organisms and neuronless metazoans (Fig. 2.14).

Summarizing the empirical evidence presented herein, it may be said that most of the molecular components and synaptic machinery of the neuron

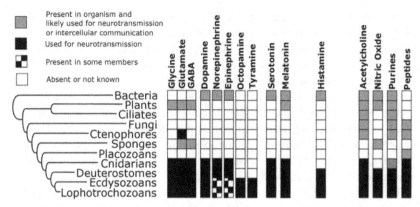

FIGURE 2.14 Evolution of neurotransmitter types. Biogenic amines derived from tyrosine, tryptophan, and histidine have a restricted taxon distribution and may be present in bilaterians + cnidarians as a result of late horizontal gene transfer of key enzymes from bacteria (Kunert and Weiser, 2003; Kunert et al., 2005). Other neurotransmitter types are used for intercellular signaling across a wider range of eukaryotes. Presence—absence data are derived from Kass—Simon and Pierobon, 2007; Elliott and Leys, 2010; Roshchina, 2010; Ruggieri et al., 2004. *From Liebeskind, B.J., Hofmann, H.A., Hillis, D.M., Zakon, H.H., 2017. Evolution of animal neural systems. Annu. Rev. Ecol. Evol. Syst. 48, 1 377—398.*

existed in neuronless metazoans. Even with that incomplete repertoire of synaptic components, they can perform several eumetazoan behaviors (directed ciliary locomotion, search for food, phototactic behavior, prey catching, etc.). This indicates that these nerveless animals possessed a *sui generis* control system for determining their behavioral adaptive responses to environmental stimuli. Hence the question is lingering:

If extant sponges are descendants of the Ediacaran sponge-grade organisms in possession of the genetic machinery and cell adhesion molecules, why sponges remained static and didn't diversify as eumetazoans did (Nichols et al., 2006). Why didn't they succeed in differentiating neurons during more than 600 million years ever since?

Fossil record and evolution of the neuron

Emergence of neuron represents a crucial event in the evolution of the Kingdom Animalia that cannot be overestimated. Among extant animals, neurons appear first in cnidarians/ctenophores, but cnidarian-grade animals are believed to have evolved during the Ediacaran and probably earlier (Ryan and Grant, 2009). An early Cambrian (~530 Ma), cnidarian-like fossil of *Xianguangia sinica* with a radially symmetrical polypoid structure is found recently in the Chengjiang fossil Lagerstätte. This suggests that the radial symmetry of *X. sinica* was ancestral to extant cnidarians and is conserved in extant hydrozoans, while is modified to biradial and bilateral body symmetry in some anthozoans (Ou et al., 2017).

The properties of ctenophore neurons, characterized by the absence of majority of the low molecular weight transmitters (acetylcholine, serotonin, GABA, dopamine, octopamine, and histamine) and the development and diversification of the glutamate, led many investigators to the idea that neurons evolved independently in ctenophora and cnidarian/bilaterians (Moroz and Kohn, 2016). According to this "ctenophora-first" hypothesis, the neuron and the nervous system arose first in the last common ancestor of all animals but were lost later in placozoans and sponges, or they evolved later two times independently (Ryan, 2014) (Fig. 2.15).

Cnidaria

Cnidaria is an animal phylum consisting of about 9000 species (Steele et al., 2011). They are diploblastic, almost all aquatic, radially symmetrical animals. Consequently, they have no right/left or dorsal/ventral sides. However, it is reported that some cnidarians at the beginning of their life cycle show some signs of bilaterality, which would support the bilaterality preceded the cnidarian−bilateralian divergence (Boero et al., 2007). Cnidarians evolved about 570 Ma, i.e., before the Cambrian explosion (Zakon, 2011).

Their body has two one-cell-thick layers of epithelial cells, the epiderm lining the outside of the body and gastroderm—the inside, developing, respectively, from embryonic layers ectoderm and gastroderm. Both layers

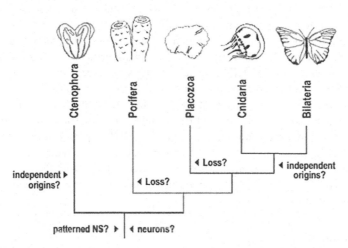

FIGURE 2.15 Phylogenetic relationship of the five main branches of animals based on recent phylogenies (Dunn et al., 2008; Hejnol et al., 2009; Ryan et al., 2013; Moroz et al., 2014). Neurons either originated before the last common ancestor of all animals and were subsequently lost in the Porifera and Placozoa lineages or arose independently at least twice. Likewise, the nervous systems (NS) either arose before the last common animal ancestor or separate NS patterning mechanisms arose independently. *From Ryan, J.F., 2016. Did the ctenophore nervous system evolve independently? Zoology (Jena) 117, 225–226.*

contain epitheliomuscular cells. Between the two layers is a noncellular layer of mesoglea, a gelatinous substance consisting of collagen, which in many species contains muscle and nerve fibers and serves as hydrostatic skeleton. In vicinity of the muscle cells, cnidarians have neurons and, in many species, cnidocytes, stinging toxin-containing cell organelles that penetrate the prey tissues or catch it. The thin body walls allow cnidarians to transport nutritive substances throughout the body via diffusion.

Cnidarians fold inward tentacles to take in the cnidocyte-paralyzed prey or suspended food particles. They have no anus and use the mouth both for taking in the food and to pump out the food waste after digestion. Food digestion involves release of digestion enzymes by the gastrodermal gland cells in the body cavity. The enzymatic treatment transforms the food in tiny suspended particles absorbable by the gastroderm cells. In some species, the endosymbionts are involved in the extracellular digestion. Cnidaria also exchange oxygen and carbon dioxide with the surrounding water via diffusion. Excretion of the nitrogen waste also is based on diffusion.

Reproduction

Cnidarian oocytes in gonads of two hydrozoan jellyfish model species, *Clytia hemisphaerica* and *Cladonema pacificum*, remain at the first meiotic prophase stage until external stimuli, light dark, and reverse transitions are sensed by the sensory organs and perceived in the nervous system. In response to these stimuli, the nervous system secretes W/RPRPamide-related neuropeptides, which stimulate some neural-type cells in the gonads to secrete MIH (maturation-inducing hormones), which act on the oocyte surface triggering oocyte maturation (Takeda et al., 2017). It is demonstrated that PRPamide is the active component of MIH and that "MIH is produced by neurosecretory cells in the gonad ectoderm" (Takeda et al., 2018). Based on the similar distribution of MIH-producing cells, it was hypothesized that MIH may also have a role in the sperm maturation and this turned out to be true in experiments where very low concentrations of synthetic MIH peptides stimulated the release of active sperm in the above cnidarian species (Takeda et al., 2018). These and other experiments have shown, beyond doubt, that the maturation and release of eggs and sperm in cnidarians is controlled and regulated by the nervous system.

Development

The cnidarians' life cycle is very complex and varies considerably among different species. It includes a medusa and a polyp stage. Most medusae have separate sexes. They release eggs/sperms in the water. After egg fecundation, the zygote develops and hatches out a $\sim 10,000$ cell large free-swimming planula, which attaches to a solid substrate. An environmental cue, mostly a bacterial component, is perceived by neurons in the aboral pole of the body (Müller and Leitz, 2002; Watanabe et al., 2009) and neurosecretory cells in the anterior part of the planula send signals that induce metamorphosis to develop into a polyp. It is demonstrated that the neurotransmitter/neuromodulator serotonin is involved in the perception of the cue (Zega et al., 2007). In the process of asexual reproduction, the polyp breaks off strobila to develop into swimming ephyra that grows to a medusa to complete the life cycle. Metamorphosis of planula into polyp is neurally regulated by several neuropeptides released by secretory neurons, among which the neuropeptide "head-inducing morphogen" (Schaller and Bodenmuller, 1981) and the neuropeptide metamorphosin A (Leitz et al., 1994). Most cnidarians lack classical neurotransmitters, but they possess a large set of neuropeptides.

Morphogenesis in cnidarians is controlled and regulated by peptide and nonpeptide substances released by neurons (Holstein et al., 1986; Schaller et al., 1996; Holstein and David, 1990). Neurons in *Hydra* secrete one small neuropeptide that induces head formation and the other foot formation. Similarly, *Hydra* possesses one neuropeptide for inhibiting the head formation and another for inhibiting foot formation. Thus, the nervous net determines pattern formation in *Hydra*.

Neuropeptide HA (head activator) and other neuropeptides commit the interstitial stem cells into the neuron pathway and differentiation into nerve cells, whereas HA secretion commits epithelial cells of the gastric region into head-specific tentacle and hypostomal cells (Schaller et al., 1996; Hampe et al., 1999). The mediator of the actions of the neuropeptide HA is a specific receptor, HAB (HA-binding protein), which is strongly expressed in the head and foot regions (Hampe et al., 1999).

Certainly, cnidarians are more complex than nerveless animals, placozoans, and sponges.

The genome

The starlet sea anemone *Nematostella vectensis* (class *Anthozoa*, phylum *Cnidaria*, a sister group of Bilateria) may be another "living fossil". Recent sequencing of its complex genome has shown that it has an estimated complement of 18,000 protein-coding genes. Its repertoire, structure, and organization is very conserved when compared with that of vertebrates but surprisingly different from that of fruit flies and nematodes, which have lost many genes and introns and have experienced genome rearrangements, indicating the genome of their common ancestor also was a complex genome (Putnam et al., 2007). The similarity of exons—intron structure of *N. vectensis* and vertebrates indicates that their common ancestor's eumetazoan genome has been intron-rich. Putative *Hox* genes in *N. vectensis* show a spatial pattern of expression that suggests a role of theirs in the embryonic development of their common ancestor. Based on their studies on the *Nematostella* genome and its conserved features in extant eumetazoans, Putnam et al. (2007) have described the common ancestor of cnidarians and vertebrates as an organism with "flagellated sperm, development through a process of gastrulation, multiple germ layers, true epithelia lying upon a basement membrane, a lined gut (enteron), a neuromuscular system, multiple sensory systems, and fixed body axes" (Putnam et al., 2007).

It is suggested that alternative splicing in cnidarians (*Hydra* and *Nematostella*) occurs, but its impact on the proteome in cnidarians is unclear (Steele et al., 2011). A clear correlation is observed, in the course of the metazoan evolution, between the proportion of the genes regulated by alternative splicing and the increase in size of the CNS.

Phylogeny of cnidarians

The phylogeny of cnidarians is not settled. The prevailing opinion now is that cnidarians and bilateria are sister groups evolved from a common ancestor, the Urmetazoa (Boero et al., 2007) (Fig. 2.16).

Fossil record

Solitary and colonial cnidarians dominated the late pre-Cambrian life (Conway Morris, 1993a,b). Fossils of microscopic cnidarian-like eumetazoans in

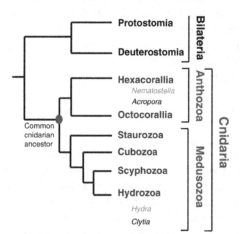

FIGURE 2.16 Phylogenetic relationships of classes in the phylum *Cnidaria*. A phylogenetic tree (based on the results of Collins, 2002 and Collins et al., 2006) showing the relationships within the phylum *Cnidaria*. The two main divisions of *Cnidaria* (*Anthozoa* and *Medusozoa*) are indicated in red (black in print version). *Anthozoa* is a class that contains two subclasses (green (gray in print version)), whereas *Medusozoa* is a subphylum consisting of four classes (green (gray in print version)). Sequenced genomes (pink (light gray in print version)) are available for *Nematostella* and *Hydra* (Chapman et al., 2010; Putnam et al., 2007), whereas the genomes of *Acropora* and *Clytia* are currently being sequenced (black). *From Technau, U., Steele, R.E., 2011. Evolutionary crossroads in developmental biology: Cnidaria. Development 138, 1447–1458.*

Doushantuo Lagerstätte are dated 580 Ma. They are represented by fossils resembling anthozoan polyps, gastrula-stage hydrozoan cnidarians, etc. (Chen et al., 2002). Fossils of medusae and medusoids are dated to Cambrian and Ordovician (Young and Hagadorn, 2010). Chitinous tubular fossils resembling cnidarians are found in the Three Gorges Ares, South China (Chang et al., 2018). A reconstruction of a lower Cambrian cnidarian-grade fossil, *X. sinica*, earlier classified as hemichordates *Chengjiangopenna wangii* and *Galeaplumosus abilus*, and its phylogenetic position is presented in Fig. 2.17.

Evolution

Cnidarians are the simplest group of the extant eumetazoans, animals with nervous system. Some biologists believe that the last common ancestor of cnidarians and bilaterians may have possessed a full-fledged nervous system, and the Na_v channels of cnidarians and bilaterians evolved from their common ancestor's Ca_v (Ca^{2+}) channels (Kelava et al., 2015).

The neural net of cnidarians exemplified in Fig. 2.18 is of the diffuse type: neurons are situated in the external and internal layers of the epithelial cells along the anterior—posterior axis and extend processes for communicating with other neurons and the muscle cells. Neurons develop not only from the

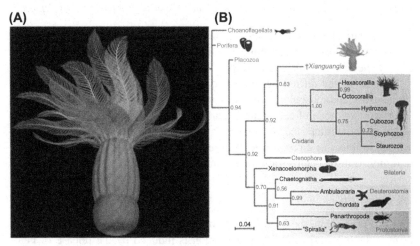

FIGURE 2.17 Reconstruction and phylogenetic position of *Xianguangla sinica*. (A) Three-dimensional model generated with 3ds Max. (B) Summary of metazoan relationships inferred from Bayesian analyses based on 111 characters and 37 taxa under Mkv + Γmodel. Numbers at nodes indicate posterior probabilities. *X. sinica* is resolved as a stem-group cnidarian. Eumetazoa, Neuralia, Bilateria, Nephrozoa, Protostomia, and Deuterostomia are monophyletic, whereas the monophyly of *Spiralia* is unresolved. Animal silhouettes are by courtesy of PhyloPic (www.phylopic.org). *From Ou, Q., Han, J., Zhang, Z., Shu, D., Sun, G., Mayer, G., 2017. Three Cambrian fossils assembled into an extinct body plan of cnidarian affinity. Proc. Natl. Acad. Sci. U.S.A. 114, 8835–8840.*

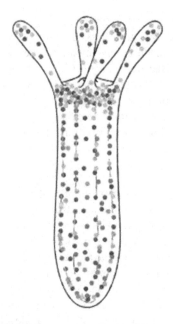

FIGURE 2.18   Distribution of different neuronal subpopulations during the development of *Nematostella vectensis. From Kelava, I., Rentzsch, F., Technau, U., 2015. Evolution of eumetazoan nervous systems: insights from cnidarians. Philos. Trans. R. Soc. Lond. B Biol. Sci. 370, 20150065.*

ectoderm but also from the endoderm. The first neuronal progenitor cells appear at midblastula stage, but formation of full-fledged neurons take place later by the late gastrula stage (Kelava et al., 2015).

Neurons of the diffuse net form synaptic connections with each other and with sensory and muscular cells. They release neuropeptides with local effect. This primitive mode of cell communication may be derived from the common cnidarian—bilaterian ancestor. However, some species display concentration of neurons in linear or circular tracts/nerve rings, which nevertheless are still far from any form of centralization (Watanabe et al., 2009).

Regeneration

During regeneration, the head formation in *Hydra* is preceded by strictly local *de novo* neurogenesis and the regeneration is induced by the release of neuropeptide HA (Schaller et al., 1989). When a *Hydra* is cut into pieces, only the middle parts of the body can regenerate head and foot, whereas isolated head or foot parts cannot. When hydra is cut into two halves, the head half regenerates the foot half and the reverse, in a process that begins very fast, in about 1—3 h after the cut (Holstein et al., 2003). The regeneration is controlled by the *de novo* neurogenesis rather than existing neurons. When neurogenesis is missing, the process of head regeneration is slower and less efficient (Miljkovic-Licina et al., 2007).

All eumetazoans (including Cnidaria and Ctenophora), both extinct and extant forms, are in possession of a universal genetic toolkit (Putnam et al., 2007). But if cnidarians and bilaterians, along ctenophores, share the same genetic toolkit, inherited from their common ancestor why Cnidaria remained "living fossils"; there are about 9000 living cnidarian species, but they show very little morphological diversity and have a common body plan (Ryan et al., 2006), while the bilaterian ancestor gave rise to several million extant and extinct species, grouped ~ 30 phyla with ~ 30 different basic body plans. If the bilaterians and cnidarians have a common genetic toolkit, what is then that determined the unprecedented phenomenal diversification and increase in complexity of bilaterians, but not cnidarians, during the Cambrian radiation?

Also the fact that sponges and placozoans have the basic genetic toolkit but did not succeed in evolving into more complex metazoans indicates that they lack something critical, the "user" of the genetic toolkit for producing organisms of higher morphological, physiological, and behavioral complexity. What is that manipulated genetic toolkit to produce enormously diversified metazoan forms within a span of ~ 10 million years during the Cambrian explosion?

Triggers of Cambrian explosion: hypotheses

Why all the known Bauplaene appeared abruptly in a geological instant more than half a billion years ago? Why the diversification of body plans did not start earlier? Why almost no new body plans evolved ever since?

Various hypotheses have been developed to answer these questions. All of them generally deal with various factors that might have played a specific role or have been necessary conditions for the Cambrian explosion. However, none of them in itself and even all of them together do not provide a theoretical explanation that would be supported by empirical evidence or that could be somehow tested. At best they present factors that might have favored the natural selection but fail to provide any mechanism of the appearance of novelties in body plans and the explosive radiation in metazoan lineages.

External triggers

Oxygen accumulation

Based on the fact that oxygen is necessary to sustain extant large animals, this hypothesis assumes that the Cambrian explosion was a consequence of the accumulation of the oxygen in the Earth atmosphere as a result of the release of the oxygen by plants, initially cyanobacteria, which began more than 2 billion years ago. The atmospheric oxygen concentration in the Early Proterozoic Age (before 2.3 billion years ago) has been $<10^{-5}$ of the present atmospheric level (Pavlov and Kasting, 2002) and during the mid-Proterozoic (roughly 1.8—0.8 billion years ago) it grew to 0.1% of the present oxygen level

(Planavsky et al., 2014), which would be inadequate for supporting metazoan life because of the high metabolic rate of multicellular animals. The second rise of the oxygen level to about 4% of the present oxygen level (Zhang et al., 2015) in atmosphere that occurred more than 1 billion years ago did not do anything in promoting the evolution of life; it took about 600 million years to the advent of metazoan life. It is believed that by the end of the global glaciations, about 580 million years ago, the rising oxygen level promoted evolution of large multicellular Ediacaran organisms (Budd, 2013). Obviously, it is hard to causally relate these oxygenation events of the Earth's atmosphere with the Cambrian explosion. Proponents of the hypothesis admit that oxygen *per se* would not have caused animals to evolve but removed a barrier to evolution of large animals (Knoll and Carroll, 1999).

The oxygen level for supporting large and complex eumetazoans existed more than 20 million years before the Cambrian (Colpron et al., 2002) and the fact that large animals appear during Ediacaran shows that there was enough oxygen to support eumetazoan life (Knoll and Carroll, 1999; Marshall, 2006) and makes it difficult to believe that oxygen or its increase was causal to the "explosion" because it cannot explain why the "explosion" had to wait 40 million years to occur.

The Snowball Earth

The temporal coincidence of the appearance of the continuous fossil record with the end of freezes of Marinoan ($\sim$635 Ma) and Gaskiers ($\sim$580 Ma) led biologists to the hypothesis of the "Snowball Earth" (Hoffman et al., 1998), which posits that as a result of the Earth's thermal subsidence for various unknown reasons (decline in the CO_2 level, reduced concentration of greenhouse gases, such as methane and/or carbon dioxide, perturbations of Earth's orbit, etc.), about 900–1000 Ma led to formation of a snow and ice cover. This is believed to have occurred between 2 and 5 times. The albedo effect (reflection of sun energy back to space) reinforced the global glaciation. The snow covering melted down soon after the level of the atmospheric carbon dioxide raised about 30 times the present level. The resulting extreme greenhouse conditions had a strong selective pressure on the evolution of life in the Proterozoic. Repeated snowball glaciations caused a series of genetic "bottleneck and flush" cycles before the advent of Ediacaran biota (Hofman et al., 1998; Marshall, 2006). But, remember, the Snowball Earth ended about 100 Ma before the Cambrian explosion and the lag of 30–40 million years between the Gaskiers glaciation and Cambrian explosion is to long to ignore, hence biologists still wonder "Why did the radiation not occur 100 million years earlier, or 100 million years later?" (Marshall, 2006).

There is evidence that a mass extinction of the Ediacaran fauna occurred during the period of transition from Ediacaran to Cambrian. However, the notion that the Cambrian radiation may be a normal postextinction biotic recovery is unwarranted. While it is true that examples of postextinction biotic recoveries of metazoans do exist, "none of the Phanerozoic mass extinctions show the sort of high-level morphological innovation seen during the Metazoan radiation" (Erwin, 2005).

Change in sea salinity

Pre-Cambrian ocean's salinity was 1.5—2 times higher than the present value (Knauth, 1998), which makes the life of higher metazoans impossible. However, when the ocean flooded continental lowlands, it created large shallow bodies of water that, over time (in an estimated 100 million years), evaporated creating salt basins. This caused a sharp fall in the ocean salinity, increased the ocean's capability to absorb oxygen, and enabled the life of aerobic life and animals with high metabolic rates. This led some biologists to the idea that the microbial life evolved first in freshwater environments and the Cambrian explosion may represent "movement of already evolved metazoans from nonmarine environments into the sea" (Knauth, 2005).

Change in the carbon—phosphorus ratio

According to an ecological stoichiometric hypothesis, the high C:P (carbon: phosphorus) ratio of the biomass in the pre-Ediacaran eon prevented metazoan invasion of the autotroph biomass and evolution of low C:P heterotroph metazoans. In other words, a stoichiometric constraint (the insufficient availability of the P in the biosphere and a mismatch in the elemental composition of the food) is considered to be the cause of the evolutionary stasis in the pre-Ediacaran eon and a later change in these factors was the cause of the Cambrian radiation (Elser et al., 2006).

Ecological hypotheses

Advent of predation

According to this hypothesis, the appearance of motile target-seeking predators would lead to a predator-prey arms race (Dawkins and Krebs, 1979; Phoenix, 2009) and a stronger selection leading to acceleration of morphological diversification of both predators and preys in direction of increasing the body mass, defensive structures like biomineralized shells, etc. (Zhang et al., 2014). The rise of interanimal predation in the early Cambrian probably provided selective pressure to evolve complex behaviors, new forms of neural organization (Phoenix, 2009; Monk and Paulin, 2014) and defensive biomineralized structures, e.g., shells in preys such as mollusks (Palmer, 1992). However, studies on the phanerozoic marine invertebrates do not support the idea (Madin et al., 2006), and latter studies, to the contrary, have shown a "negative correlation coefficient" suggesting that "predators may have impeded the Cambrian—Ordovician radiation" (López-Villalta, 2016).

Like other "detonators," arms race may have played selective role, but it might not accelerate in any imaginable way the inherent propensity of animals for evolutionary change.

Bioturbation

The accumulation of microorganisms (bacteria, archaea, and fungi) for more than 2 billion years before the Ediacaran era formed a multilayered microbial mat with one-dimensional structure on the sea bed. There was little interaction between the biogenic sediment and the sea water. Macroscopic animals during the Ediacaran feed on the organic matter or on other animals on the microbial mats. The mat started disappearing about 40−60 million years before the Cambrian explosion (Meysman et al., 2006). With the appearance of predators, the prey used burrowing within the deeper layers of the microbially bound substrate (matground) to avoid or escape predators. This led to bioturbation down to the sediment depth (Seilacher and Pflüger, 1994; Bottjer et al., 2000) and destruction of the wide blanket of microbial mats, transforming the matground into a mixground.

The burrowing traces in the biogenic matground were generally horizontal, simple and rare, and without tracks of arthropods or sinusoidal nematodes. This led to the so-called "agronomic revolution" (Seilacher and Pflüger, 1994) or the Cambrian substrate revolution (Bottjer et al., 2000).

During the Ediacaran−Cambrian transition, suddenly in the microbial mats appear vertical and complex burrowing traces (Seilacher et al., 2005), suggesting emergence of animals with higher neural processing and integration of environmental signals and rapid increase of the Cambrian ichnofauna. This destroyed the sediment strata and modified sediment chemical and physical properties leading to the phenomenon of bioturbation as a possible trigger of the Cambrian explosion (Knoll and Carroll, 1999) (Fig. 2.19).

All this may be loyally describing the real bioecological occurrences of the period, but while the resulting changes in the habitat created new niches and selection pressures, it is hard to understand how the agronomic revolution would accelerate the rate of the evolutionary change that led to the explosive diversification of animal forms after the Ediacaran.

Extinction of Ediacaran biota

No targeted interanimal predation is believed to have been practiced during the Ediacaran (Phoenix, 2009), but the situation changed suddenly on the eve of the Cambrian with the advent of predator animals. The extinction of the Ediacaran biota by the beginning of the Cambrian created an empty ecosystem for the Cambrian biota to flourish (Seilacher, 1984; Narbonne, 2005). At the outset of the Cambrian, the number of the Ediacaran survivor species "can be counted on the fingers of one hand" (Peterson et al., 2003). At a smaller scale, a similar event is recorded during the Cretaceous−Paleogene ($\sim$ 66 Ma), when

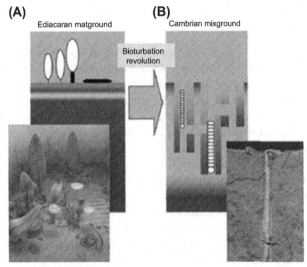

FIGURE 2.19 Transition from Ediacaran matground to Cambrian mixground during the burrowing revolution. During the Ediacaran, sediments were covered with microbial mats (A). After the burrowing revolution, the sediment is mixed and intersected by complex burrow networks (B). As a result of burrow flushing, oxygen is now transported deep into the sediment. The lugworm *Arenicola marina* has a light halo of oxidized sediment around its burrow as opposed to the gray background of reduced sediment. *(A). (Vendian diorama, photo reproduced with permission from William Hargrove). (B). (photo reproduced with permission from Oleksiy Galaktionov). Adapted with permission from Seilacher, A., Pflüger, F., 1994. From biomats to benthic agriculture: a biohistoric revolution. In: (Krumbein, W.E. et al., Eds.), Biostabilization of Sediments Universität Oldenburg 97—105. From Meysman, F.J.R., Middelburg, J.J., Heip, C.H.R., 2006. Bioturbation: a fresh look at Darwin's last idea. Trends Ecol. Evol. 21, 688—695.*

the elimination of almost all the large-body animals made the resources they utilized available to the surviving species and opened for them new niches leading, thus, to the well-known great morphological-taxonomic diversification of the classes of mammals and birds (Losos, 2010).

Neither of the processes mentioned above as triggers of the Cambrian explosion could explain the abruptness of the Cambrian radiation of animal forms. None of the above agents can produce adaptive changes in gene expression. What they do instead is to create new selection pressures for or against evolution in particular directions rather than create or increase the probability of heritable changes at the individual level, which is the source of the evolutionary change in metazoans like other forms of life on earth. But first and at the heart of the problem of evolution, is the mechanism of the emergence of the evolutionary rather than its selection process, which is satisfactorily elaborated during these 160 years since the publication of *The Origin*.

Internal triggers

Evolution of Hox genes

This genetic hypothesis posits that the Cambrian explosion was triggered by evolution of regulatory or "master" *Hox* genes (Erwin and Valentine, 2013), which control expression of other genes. It is suggested that some modifications of the *Hox* assemblage and other regulatory changes downstream might have taken place before the Cambrian explosion could happen (Valentine et al., 1999), but there is no evidence and plausible explanation why it might have occurred. The hypothesis is rejected by the evidence that the *Hox* family of genes predates the Cambrian explosion (de Rosa et al., 1999), i.e., it was present in metazoans about 50–100 million years before the Cambrian explosion. Moreover, a comparative review shows that often animal complexity is associated with the decrease in the number of Hox genes. So, e.g., *Drosophila*, which is by far a more complex organism than *N. vectensis*, a cnidarian species of the class *Anthozoa*, has five times more cell types, but much less *Hox* genes (Ryan et al., 2006).

The advent of the developmental mechanisms of control

According to this hypothesis, for the evolution of new body plans, only new gene regulatory networks (GRNs) rather than additions in the existing gene repertoire were necessary (Peterson and Davidson, 2000). GRN rewiring via cooption of GRN subcircuits is the most common mechanism of developmental processes that generate upper-level evolutionary changes in the offspring (Peter and Davidson, 2011). The hypothesis also stresses the role of noncoding miRNAs as a new layer of the control of protein biosynthesis. It is suggested that the reassembly and cooption of the GRN subcircuits in other regions of the body may be responsible for the explosive diversification of metazoans during the Cambrian (Zhang et al., 2014). The evolution of new body plans was combined with the predation, triploblasty, and the evolution of complex life cycles (biphasic life cycle, comprising a larval stage) (Peterson et al., 2005).

Still extant bilateria retain a relatively similar toolkit of regulatory genes (Erwin and Davidson, 2002), and the developmentally relevant dGRNs (developmental gene regulatory networks) presently shared among metazoan lineages were in place in Urbilateria, about 555 million years ago (Zhang et al., 2014), but a crucial process that evolved during the Cambrian explosion is the accelerated incorporation of miRNA dGRNs (Erwin et al., 2011) as regulators of mRNA translation. What essentially changed instead with the evolution of bilaterians was elaboration of patterns of developmental control and regulation (Davidson and Erwin, 2006).

The hypothesis marks considerable progress to understanding causal basis of the Cambrian by bringing the inherent animal evolutionary potential into the driver's seat of the evolution. However, the hypothesis fails to notice and appreciate that almost all of the GRNs described in detail are triggered by maternal cytoplasmic factors, which clearly are epigenetic rather than genetic factors. It also fails to explain two important facts:

First, why it took so long for the Cambrian to occur when the toolkit of regulatory genes and dGRNs in bilaterians has been in place at least tens of millions of years earlier.

Second, it seems difficult to reconcile the explosive Cambrian radiation with the high conservation of the dGRNs.

References

Antcliffe, J.B., 2013. Questioning the evidence of organic compounds called sponge biomarkers. Palaeontology 56, 917–925.

Antcliffe, J.B., Callow, R.H., Brasier, M.D., 2014. Giving the early fossil record of sponges a squeeze. Biol. Rev. Camb. Philos. Soc. 89 (2014), 972–1004.

Arabzadeh, E., Panzeri, S., Diamond, M.E., 2006. Deciphering the spike train of a sensory neuron: counts and temporal patterns in the rat whisker pathway. J. Neurosci. 26, 9216–9226.

Bengtson, S., Budd, G., 2004. Comment on small bilaterian fossils from 40 to 55 million years before the Cambrian. Science 306, 1291.

Bobrovskiy, I., Hope, J.M., Ivantsov, A., Nettersheim, B.J., Hallmann, C., Brocks, J.J., 2018. Ancient steroids establish the Ediacaran fossil *Dickinsonia* as one of the earliest animals. Science 361, 1246–1249.

Boero, F., Schierwater, B., Piraino, S., 2007. Cnidarian milestones in metazoan evolution. Integr. Comp. Biol. 47, 693–700.

Bottjer, D.J., Hagadorn, J.W., Dornbos, S.Q., 2000. The Cambrian substrate revolution. GSA Today (Geol. Soc. Am.) 10, 2–7.

Botting, J.P., Cárdenas, P., Peel, J.S., 2015. A crown-group demosponge from the early Cambrian Sirius Passet biota, North Greenland. Palaeontology 58, 35–43.

Botting, J.P., Muir, L.A., 2017. Early sponge evolution: a review and phylogenetic framework. Palaeoworld 27, 1–29.

Brain, C., Prave, A.R., Hoffman, K.-H., Fallick, A.E., Botha, A., Herd, D.A., Sturrock, C., Young, I., Condon, D.J., Allison, S.G., 2012. The first animals: ca. 760-million-year-old sponge-like fossils from Namibia. South Afr. J. Sci. 108, 83–90.

Brunet, T., Arendt, D., 2016. From damage response to action potentials: early evolution of neuronal and contractile modules in stem eukaryotes. Philos. Trans. R. Soc. Lond. Biol. Sci. 371, 20150043.

Budd, G.E., 2008. The earliest fossil record of the animals and its significance. Philos. Trans. R. Soc. B 363, 1425–1434.

Budd, G.E., 2013. At the origin of animals: the revolutionary Cambrian fossil record. Curr. Genom. 14, 344–354.

Budd, G.E., Jensen, S., 2017. The origin of the animals and a 'Savannah' hypothesis for early bilaterian evolution. Biol. Rev. 92, 446–473.

Burkhardt, P., Sprecher, S.G., 2017. Evolutionary origin of synapses and neurons – bridging the gap. Bioessays 39 (10), 1700024.

Butterfield, N.J., 2006. Hooking some stem-group "worms": fossil lophotrochozoans in the Burgess Shale. Bioessays 28, 1161–1166.

Butterfield, N.J., 2008. An early Cambrian *Radula*. J. Paleontol. 82 (3), 543—554.

Cavalier-Smith, T., 2017. Origin of animal multicellularity: precursors, causes, consequences—the choanoflagellate/sponge transition, neurogenesis and the Cambrian explosion. Phil. Trans. R. Soc. B 372, 20150476.

Chang, S., Feng, Q., Clausen, S., Zhang, L., 2017. Sponge spicules from the lower Cambrian in the Yanjiahe Formation, South China: the earliest biomineralizing sponge record. Palaeogeogr. Palaeoclimatol. Palaeoecol. 474, 36—44.

Chang, S., Clausen, S., Zhang, L., Feng, Q., Steiner, M., Bottjer, D.J., et al., 2018. New probable cnidarian fossils from the lower Cambrian of the Three Gorges area, South China, and their ecological implications. Palaeogeogr. Palaeoclimatol. Palaeoecol. 505, 150—166.

Chen, J.-Y., Bottjer, D.J., Oliveri, P., Dornbos, S.Q., Gao, F., Ruffins, S., et al., 2004. Small bilaterian fossils from 40 to 55 million years before the Cambrian. Science 305, 218—222.

Chen, J.-Y., Oliveri, P., Gao, F., Dornbos, S.Q., Li, C.-W., Bottjer, D.J., Davidson, E.H., 2002. Precambrian animal life: probable developmental and adult cnidarian forms from southwest China. Dev. Biol. 248, 182—196.

Chapman, J.A., Kirkness, E.F., Simakov, O., Hampson, S.E., Mitros, T., Weinmaier, T., et al., 2010. The dynamic genome of *Hydra*. Nature 464, 592—596.

Colpron, M., Logan, J.M., Mortensen, J.K., 2002. U-Pb zircon age constraint for late Neo-proterozoic rifting and initiation of the lower Paleozoic passive margin of western Laurentia. Can. J. Earth Sci. 39, 133—143.

Collins, A.G., 2002. Phylogeny of Medusozoa and the evolution of cnidarian life cycles. J. Evol. Biol. 15, 418—432.

Collins, A.G., Bentlage, B., Matsumoto, G.I., Haddock, S.H., Osborn, K.J., Schierwater, B., 2006. Solution to the phylogenetic enigma of *Tetraplatia*, a worm-shaped cnidarian. Biol. Lett. 2, 120—124.

Conaco, C., Bassett, D.S., Zhou, H., Arcila, M.L., Degnan, S.M., Degnan, B.M., et al., 2012. Functionalization of a protosynaptic gene expression network. Proc. Natl. Acad. Sci. U.S.A. 109, 10612—10618.

Conway Morris, S., 1993a. Ediacaran-like fossils in Cambrian Burgess Shale-type faunas of north America. Palaeontology 36, 593—635.

Conway Morris, S., 1993b. The fossil record and the early evolution of the metazoa. Nature 361, 219—225.

Cunningham, J.A., Vargas, K., Pengju, L., Belivanova, V., Marone, F., Martínez-Pérez, C., et al., 2015. Critical appraisal of tubular putative eumetazoans from the Ediacaran Weng'an Doushantuo biota. Proc. R. Soc. B 282, 20151169.

Darroch, S.A., Sperling Boag, T.H., Racicot, R.A., Mason, S.J., Morgan, A.S., et al., 2015. Biotic replacement and mass extinction of the Ediacara biota. Proc. R. Soc. B 282, 20151003.

Darroch, S.A., Boag, T.H., Racicot, R.A., Tweedt, S., Mason, S.J., Erwin, D.H., et al., 2016. A mixed Ediacaran-metazoan assemblage from the Zaris Sub-basin, Namibia. Palaeogeogr. Palaeoclimatol. Palaeoecol. 459, 198—208.

Davidson, E.H., Erwin, D.H., 2006. Gene regulatory networks and the evolution of animal body plans. Science 311, 796—800.

Dawkins, R., Krebs, J.R., 1979. Arms races between and within species. Proc. Roy. Soc. Lond. B Biol. Sci. 205, 489—511.

de Rosa, R., Grenier, J.K., Andreeva, T., Cook, C.E., Adoutte, A., Akam, M., Carroll, S.B., Balavoine, G., 1999. *Hox* genes in brachiopods and priapulids and protostome evolution. Nature 399, 772—776.

dos Reis, M., et al., 2015. Uncertainty in the timing of origin of animals and the limits of precision in molecular timescales. Curr. Biol. 25, 2939–2950.

Droser, M.L., Tarhan, L.G., Gehling, J.G., 2017. The rise of animals in a changing environment: global ecological innovation in the late Ediacaran. Annu. Rev. Earth Planet Sci. 45, 593–617.

Dunn, C.W., Hejnol, A., Matus, D.Q., Pang, K., Browne, W.E., Smith, S.A., et al., 2008. Broad phylogenomic sampling improves resolution of the animal tree of life. Nature 452, 745–749.

Dunn, F.S., Liu, A.G., 2017. Fossil focus: the Ediacaran biota. Palaeontology Online 7 (1), 1–15. Internet. http://www.palaeontologyonline.com/articles/2017/fossil-focus-ediacaran-biota.

Elliott, G.R.D., Leys, S.P., 2007. Coordinated contractions effectively expel water from the aquiferous system of a freshwater sponge. J. Exp. Biol. 210, 3736–3748.

Elliott, G.R.D., Leys, S.P., 2010. Evidence for glutamate, GABA and NO in coordinating behaviour in the sponge, Ephydatia muelleri Demospongiae, Spongillidae. J. Exp. Biol. 213, 2310–2321.

Ellwanger, K., Nickel, M., 2006. Neuroactive substances specifically modulate rhythmic body contractions in the nerveless metazoan *Tethya wilhelma* (Demospongiae, Porifera). Front. Zool. 3 (2006), 7.

Elser, J.J., Watts, J., Schampell, J.H., Farmer, J., 2006. Early Cambrian food webs on a trophic knife-edge? Ecol. Lett. 9, 292–300.

Emes, R., Pocklington, A.J., Anderson, C.N.G., Bayes, A., Collins, M.O., Vickers, C.A., et al., 2008. Evolutionary expansion and anatomical specialization of synapse proteome complexity. Nature Neurosci 11, 799–806.

Emes, R.D., Grant, S.G.N., 2012. Evolution of synapse complexity and diversity. Annu. Rev. Neurosci. 35, 111–131.

Erwin, D.H., 2005. The origin of animal body plans. In: Briggs, D.E.G. (Ed.), Evolving Form and Function: Fossils and Development. Proceedings of a Symposium Honoring Adolf Seilacher for His Contributions to Paleontology, in Celebration of His 80th Birthday. Peabody Museum. April 1–2, 2005.

Erwin, D.H., 2015. Early metazoan life: divergence, environment and ecology. Philos. Trans. R. Soc. Lond. B Biol. Sci. 370, 20150036.

Erwin, D.H., Davidson, E.H., 2002. The last common bilaterian ancestor. Development 129, 3021–3032.

Erwin, D.H., Laflamme, M., Tweedt, S.M., Sperling, E.A., Pisani, D., Peterson, K.J., 2011. The Cambrian conundrum: early divergence and later ecological success in the early history of animals. Science 334, 1091–1097.

Erwin, D.H., Valentine, J.W., 2013. The Cambrian Explosion. Roberts and Company Publishers, Greenwood Village Co, p. 261.

Evans, S.D., Diamond, C.W., Droser, M.L., Lyons, T.W., 2018. Dynamic oxygen and coupled biological and ecological innovation during the second wave of the Ediacara Biota. Emerg. Topics Life Sci. 2, 223–233.

Fedonkin, M.A., Waggoner, B.M., 1997. The late Precambrian fossil *Kimberella* is a mollusc-like bilaterian organism. Science 388, 868–871.

Fedonkin, M.A., Vickers-Rich, P., 2007. First trace of motion. In: Fedonkin, M.A., Gehling, J.G., Grey, K., Narbonne, G.M., Vickers-Rich, P. (Eds.), The Rise of Animals. Johns Hopkins University Press, Baltimore, pp. 205–216.

Fedonkin, M.A., Simonetta, A., Ivantsov, A.Y., 2007. New Data on *Kimberella*, the Vendian Mollusc-like Organism (White Sea Region, Russia): Palaeoecological and Evolutionary Implications, vol. 286. Geological Society, London, Special Publications, pp. 157–179.

Feldman, J., 2010. Ecological expected utility and the mythical neural code. Cogn. Neurodyn. 4, 25−35.

Gehling, J.G., Droser, M.L., Jensen, S., Runnegar, B.N., 2005. Ediacaran organisms: relating form to function. In: Briggs, D.E.G. (Ed.), Evolving Form and Function: Fossils and Development, Proceedings of a Symposium Honoring Adolf Seilacher for His Contributions to Palaeontology in Celebration of His 80th Birthday: New Haven, Connecticut, Peabody Museum of Natural History. Yale University, pp. 43−67.

Glaesner, M.F., Wade, M., 1966. The late precambrian fossils from Ediacara, south Australia. Palaeontology 9, 599−628.

Gold, D.A., Grabenstatter, J., de Mendoza, A., Riesgo, A., Ruiz-Trillo, I., Summons, R.E., 2016. Sterol and genomic analyses validate the sponge biomarker hypothesis. Proc. Natl. Acad. Sci. U.S.A. 113, 2684−2689.

Hampe, et al., 1999. A head-activator binding protein is present in *Hydra* in a soluble and a membrane-anchored form. Development 126, 4077−4086.

Hejnol, A., Obst, M., Stamatakis, A., Ott, M., Rouse, G.W., Edgecombe, G.D., et al., 2009. Assessing the root of bilaterian animals with scalable phylogenomic methods. Proc. R. Soc. Lond. B 276, 4261−4270.

Hoekzema, R.S., Brasier, M.D., Dunn, F.S., Liu, A.G., 2017. Quantitative study of developmental biology confirms *Dickinsonia* as a metazoan. Proc. Biol. Sci. 284, 20171348.

Hoffman, P.F., Kaufman, A.J., Halverson, G.P., Schrag, D.P., 1998. A neoproterozoic snowball earth. Science 281, 1342−1346.

Holstein, T., Schaller, C.H., David, C.N., 1986. Nerve cell differentiation in *Hydra* requires two signals. Dev. Biol. 115, g−17.

Holstein, T.W., David, C.N., 1990. Putative Intermediates in the Nerve Cell Differentiation Pathway in *Hydra* Have Properties of Multipotent Stem Cells. Dev. Bio. 142, pp. 401−405.

Holstein, T.W., Hobmayer, E., Technau, U., 2003. Cnidarians: an evolutionarily conserved model system for regeneration? Dev. Dynam. 226, 257−267.

Ivantsov, A.Y., Malakhovskaya, Y.E., 2002. Giant traces of Vendian animals. Dokl. Earth Sci. 385 (No. 6), 618−622. Translated from Doklady Akademii Nauk, Vol. 385, No. 3, 2002, pp. 382−386.

Jacobs, D.K., Nakanishi, N., Yuan, D., Camara, A., Nichols, S.A., Hartenstein, V., et al., 2007. Evolution of sensory structures in basal metazoan. Integr. Comp. Biol. 47, 712−723.

Jékely, G., 2013. Global view of the evolution and diversity of metazoan neuropeptide signaling. Proc. Natl. Acad. Sci. U.S.A. 110, 8702−8707.

Kass−Simon, G., Pierobon, P., 2007. Cnidarian chemical neurotransmission, an updated overview. Comp. Biochem. Physiol. A. Mol. Integr. Physiol. 146, 9−25.

Kelava, I., Rentzsch, F., Technau, U., 2015. Evolution of eumetazoan nervous systems: insights from cnidarians. Philos. Trans. R. Soc. Lond. B Biol. Sci. 370, 20150065.

King, N., Hittinger, C.T., Carroll, S.B., 2003. Evolution of key cell signaling and adhesion protein families predates animal origins. Science 361−363.

King, N., Westbrook, M.J., Young, S.L., Kuo, A., Abedin, M., Chapman, J., et al., 2008. The genome of the choanoflagellate *Monosiga brevicollis* and the origin of metazoans. Nature 451, 783−788.

Knauth, L.P., 1998. Salinity history of the earth's early ocean. Nature 395, 554−555.

Knauth, L.P., 2005. Temperature and salinity history of the Precambrian ocean: implications for the course of microbial evolution. Palaeogeogr. Palaeoclimatol. Palaeoecol. 219, 53−69.

Knoll, A.H., Carroll, S., 1999. Early animal evolution: emerging views from comparative biology and geology. Science 284, 2129−2137.

Kristan Jr., W.B., 2016. Early evolution of neurons. Curr. Biol. 26, R949–R954.

Kunert, G., Weiser, W.W., 2003. The interplay between density- and trait-mediated effects in predatory-prey interactions: a case study in aphid wing polymorphism. Oecologia 135, 304–312.

Kunert, G., Otto, S., Roese, U.S.R., Gershenzon, J., Weiser, W.W., 2005. Alarm pheromone mediates production of winged dispersal morphs in aphids. Ecol. Lett 8, 596–603.

Laflamme, M., Darroch, S.A.F., Tweedt, S.M., Peterson, K.J., Erwin, D.H., 2013. Gondwana Res. 23, 558–573.

Leitz, T., Morand, K., Mann, M., 1994. A novel peptide controlling development of the lower metazoan *Hydractinia echinata* (Coelenterata, Hydrozoa). Dev. Biol. 163, 440–446.

Leys, S.P., 2015. Elements of a 'nervous system' in sponges. J. Exp. Biol. 218, 581–591.

Leys, S.P., Mackie, G.O., Meech, R.W., 1999. Impulse conduction in a sponge. J. Exp. Biol. 202, 1139–1150.

Leys, S.P., Degnan, B.M., 2001. Cytological basis of photoresponsive behavior in a sponge larva. Biol. Bull. 201, 323–338.

Leys, S.P., Cronin, T.W., Degnan, B.M., Marshall, J.N., 2002. Spectral sensitivity in a sponge larva. J. Comp. Physiol. A Neuroethol. Sens. Neural Behav. Physiol. 188, 199–202.

Leys, S.P., Eerkes-Medrano, D., 2005. Gastrulation in calcareous sponges: in search of Haeckel's gastraea. Integr. Comp. Biol. 45, 342–351.

Leys, S.P., Bernard, M., 2005. Degnan embryogenesis and metamorphosis in a haplosclerid demosponge. Invertebr. Biol. 171–189.

Leys, S., Meech, R.W., 2011. Physiology of coordination in sponges. Can. J. Zool. 84, 288–306.

Li, C.W., Chen, J.Y., Hua, T.E., 1998. Precambrian sponges with cellular structures. Science 79, 879–882.

Liebeskind, B.J., Hofmann, H.A., Hillis, D.M., Zakon, H.H., 2017. Evolution of animal neural systems. Annu. Rev. Ecol. Evol. Syst. 48 (1), 377–398.

Liu, P., Xiao, S., Yin, C., Zhou, C., GAO, L., Tang, F., 2008. Systematic description and phylogenetic affinity of tubular microfossiks from the Ediacaran Doushantua formation at Weng'anm south China. Palaeontology 51, 339–366.

López-Villalta, J.S., 2016. Testing the predation-diversification hypothesis for the Cambrian—ordovician radiation. Paleontol. Res. 20, 312–321.

Losos, J.B., 2010. Adaptive radiation, ecological opportunity, and evolutionary determinism. Am. Nat. 175, 623–639.

Love, G.D., Grosjean, E., Stalvies, C., Fike, D.A., Grotzinger, J.P., Bradley, A.S., et al., 2009. Fossil steroids record the appearance of *Demospongiae* during the Cryogenian period. Nature 457, 718–721.

Ludeman, D.A., Farrar, N., Riesgo, A., Paps, J., Leys, S.P., 2014. Evolutionary origins of sensation in metazoans: functional evidence for a new sensory organ in sponges. BMC Evol. Biol. 14 (2014), 3.

Madin, J.S., Alroy, J., Aberhan, M., Fürsich, F.T., Kiessling, W., Kosnik, M.A., Wagner, P.J., 2006. Statistical independence of escalatory ecological trends in Phanerozoic marine invertebrates. Science 312, 897–900.

Marshall, C.R., 2006. Explaining the Cambrian "explosion" of animals. Annu. Rev. Earth Planet Sci. 34, 355–384.

Martin, M.W., Grazhdankin, D.V., Bowring, S.A., Evans, D.A.D., Fedonkin, M.A., Kirschvink, J.L., 2000. Age of Neoproterozoic bilatarian body and trace fossils, White Sea, Russia: implications for metazoan evolution. Science 288, 841–845.

Meroz-Fine, E., Shefer, S., Ilan, M., 2005. Changes in morphology and physiology of an East Mediterranean sponge in different habitats. Mar. Biol. 147, 243—250.

Meysman, F.J.R., Middelburg, J.J., Heip, C.H.R., 2006. Bioturbation: a fresh look at Darwin's last idea. Trends Ecol. Evol. 21, 688—695.

Miljkovic-Licina, M., Chera, S., Ghila, L., Galliot, B., 2007. Head regeneration in wild-type *Hydra* requires *de novo* neurogenesis. Development 134, 1191—1201.

Monk, T., Paulin, M.G., 2014. Evolutionary origins of sensation in metazoans: functional evidence for a new sensory organ in sponges. Brain Behav. Evol. 84, 246—261.

Moroz, L., 2012. Phylogenomics meets neuroscience: how many times might complex brains have evolved? Acta Biol. Hung. 63 (Suppl. l), 3—19, 2.

Moroz, L.L., Kocot, K.M., Citarella, M.R., Dosung, S., Norekian, T.P., Povolotskaya, I.S., et al., 2014. The ctenophore genome and the evolutionary origins of neural systems. Nature. https://doi.org/10.1038/nature13400.

Moroz, L.L., Kohn, A.B., 2016. Independent origins of neurons and synapses: insights from ctenophores. Philos. Trans. R. Soc. Lond. B Biol. Sci. 371 (1685).

Müller, W.A., Leitz, T., 2002. Metamorphosis in the Cnidaria. Can. J. Zool. 80, 1755—1771.

Nakanishi, N., Sogabe, S., Degnan, B.M., 2014. Evolutionary origin of gastrulation: insights from sponge development. BMC Biol. 12 (2014), 26.

Nakanishi, N., Stoupin, D., Degnan, S.M., Degnan, B.M., 2015. Sensory flask cells in sponge larvae regulate metamorphosis via calcium signaling. Integr. Comp. Biol. 55, 1018—1027.

Narbonne, G.M., 2005. The Ediacara biota: Neoproterozoic origin of animals and their ecosystems. Annu. Rev. Earth Planet Sci. 33, 421—442.

Nettersheim, B.J., Brocks, J.J., Schwelm, A., Hope, J.M., Not, F., Lomas, M., et al., 2019. Putative sponge biomarkers in unicellular *Rhizaria* question an early rise of animals. Nat. Ecol. Evol. 3, 577—581.

Newman, S.A., Müller, G.B., 2001. Morphological evolution: epigenetic mechanisms. Encycl. Life Sci. Internet: https://www.researchgate.net/publication/227998057_Morphological_Evolution_Epigenetic_Mechanisms.

Newman, S.A., Müller, G.B., 2001. Epigenetic mechanisms of character origination. J. Exp. Zool. 288, 304—317.

Nichols, S.A., Dirks, W., Pearse, J.S., King, N., 2006. Early evolution of animal cell signaling and adhesion genes. Proc. Natl. Acad. Sci. U.S.A. 103, 12451—12456.

Nikitin, M., 2015. Bioinformatic prediction of *Trichoplax adhaerens* regulatory peptides. Gen. Comp. Endocrinol. 212, 145—155.

Ou, Q., Han, J., Zhang, Z., Shu, D., Sun, G., Mayer, G., 2017. Three Cambrian fossils assembled into an extinct body plan of cnidarian affinity. Proc. Natl. Acad. Sci. U.S.A. 114, 8835—8840.

Palmer, A.R., 1992. Calcification in marine molluscs: how costly is it? Proc. Natl. Acad. Sci. U.S.A. 89, 1379—1382.

Pavlov, A.A., Kasting, J.F., 2002. Mass-independent fractionation of sulbib isotopes in Archean sediments: strong evidence for an anoxic Archean atmosphere. Astrobiology 2, 27—41.

Peter, I.S., Davidson, E.H., 2011. Evolution of gene regulatory networks controlling body plan development. Cell 144, 970—985.

Peterson, K.J., Davidson, E.H., 2000. Regulatory evolution and the origin of the bilaterians. Proc. Natl. Acad. Sci. U.S.A. 97, 4430—4433.

Peterson, K.J., et al., 2004. Estimating metazoan divergence times with a molecular clock. Proc. Natl. Acad. Sci. U.S.A. 101, 6536—6541.

Peterson, K.J., McPeek, M.A., Evans, D.A.D., 2005. Tempo and mode of early animal evolution: inferences from rocks, *Hox* and molecular clocks. Paleobiology 31 (Suppl. l), 36—55.

Peterson, K.J., Dietrich, M.R., McPeek, M.A., 2009. MicroRNAs and metazoan macroevolution: insights into canalization, complexity, and the Cambrian explosion. Bioessays 31, 736–747.

Peterson, K.J., Waggoner, B., Hagedorn, J.W., 2003. A fung analog for newfoundland Ediacaran fossils? Integr. Comp. Biol. 43, 127–136.

Pflug, H.D., 1973. Zur fauna der Nama-Schichten in Südwest-Afrika. IV. Mikroscopische Anatomie der Petalo-organisme. *Paleontographica* (B144), pp. 166–202.

Phoenix, C., 2009. Cellular differentiation as a candidate "new technology" for the Cambrian Explosion. J. Evol. Technol. 20, 43–48.

Planavsky, N.J., Reinhard, C.T., Wang, X., Thomson, D., McGoldrick, P., Rainbird, R.H., et al., 2014. Low mid-proterozoic atmospheric oxygen levels and the delayed rise of animals. Science 346, 635–638.

Porter, S.M., Meisterfeld, R., Knoll, A.H., 2003. Vase-shaped microfossils from the Neoproterozoic Chuar Group, Grand Canyon: a classification guided by modern testate amoebae. J. Paleontol. 77, 409–429.

Putnam, N.H., Srivastava, M., Hellsten, U., Dirks, B., Chapman, J., Salamov, A., et al., 2007. sea anemone genome reveals ancestral eumetazoan gene repertoire and genomic organization. Science 317, 86–94.

Richards, G.S., Simionato, E., Perron, M., Adamska, M., Vervoort, M., Degnan, B.M., et al., 2008. Sponge genes provide new insight into the evolutionary origin of the neurogenic circuit. Curr. Biol. 18, 1156–1161.

Rivera, A.S., Ozturk, N., Fahey, B., Plachetzki, D.C., Degnan, B.M., Sancar, A., Oakley, T.H., 2012. Blue-light-receptive cryptochrome is expressed in a sponge eye lacking neurons and opsin. J. Exp. Biol. 215, 1278–1286.

Roshchina, V.V., 2010. Evolutionary Considerations of Neurotransmitters in Microbial, Plant, and Animal Cells. Microbial Endocrinology. In: Lyte, M., Freestone, P.P.E. (Eds.). Springer, New York, pp. 17–52.

Ruggieri, R.D., Pierobon, P., Kass–Simon, G., 2004. Pacemaker activity in hydra is modulated by glycine receptor ligands. Comp. Biochem. Physiol. A. Mol. Integr. Physiol. 138, 193–202.

Ryan, J.F., Burton, P.M., Mazza, M.E., Kwong, G.K., Mullikin, J.C., Finnerty, J.R., 2006. The cnidarian-bilaterian ancestor possessed at least 56 homeoboxes: evidence from the starlet sea anemone, *Nematostella vectensis*. Genome Biol. 7 (2006), R64, 24.

Ryan, T.J., Grant, S.G.N., 2009. The origin and evolution of synapses. Nat. Rev. Neurosci. 10, 701–712.

Ryan, J.F., Pang, K., Schnitzler, C.E., Nguyen, A.D., Moreland, R.T., Simmons, D.K., et al., 2013. The Genome of the Ctenophore Mnemiopsis leidyi and its Implications for Cell Type Evolution Science, vol. 342, p. 1242592.

Ryan, J.F., 2014. Did the ctenophore nervous system evolve independently? Zoology (Jena), 117, 225–226.

Sakarya, O., Armstrong, K.A., Adamska, M., Adamski, M., Wang, I.F., Tidor, B., Degnan, B.M., Oakley, T.H., Kosik, K.S., 2007. A post–synaptic scaffold at the origin of the animal kingdom. PLoS One 2, e506.

Schaller, H.C., Bodenmuller, H., 1981. Isolation and amino acid sequence of a morphogenetic peptide from *Hydra*. Dev. Biol. 78, 7000–7004.

Schaller, H.C., Hermans-Borgmeyer, I., Hoffmeister, S.A.H., 1996. Neuronal control of development in *Hydra*. Int. J. Dev. Biol. 40, 339–344.

Schaller, H.C., Hoffmeister, S.A., Dubel, S., 1989. Role of the neuropeptide head activator for growth and development in *Hydra* and mammals. Development 107, 99–107.

Seilacher, A., 1984. Late precambrian and early Cambrian metazoa: preservational or real extinctions? In: Holland, H.D., Trendall, A.F. (Eds.), Patterns of Change in Earth Evolution. Dahlem Workshop Reports Physical, Chemical, and Earth Sciences Research Reports, vol. 5. Springer, Berlin — Heidelberg.

Seilacher, A., 1989. Vendozoa: organismic construction in the proterozoic biosphere. Lethaia 22, 229—239.

Seilacher, A., 1992. Vendobionta and psammocorallia: lost constructions of Precambrian evolution. J. Geol. Soc. 149, 607—613. London.

Seilacher, A., Pflüger, F., 1994. From biomats to benthic agriculture: a biohistoric revolution. In: Krumbein, W.E., et al. (Eds.), Biostabilization of Sediments. Universität Oldenburg, pp. 97—105.

Seilacher, A., Buatois, L.A., Mángano, M.G., 2005. Trace fossils in the Ediacaran-Cambrian transition: behavioral diversification, ecological turnover and environmental shift. Palaeogeogr. Palaeoclimatol. Palaeoecol. 227, 323—356.

Senatore, A., Raiss, H., Le, P., 4 November, 2016. Physiology and evolution of voltage-gated calcium channels in early diverging animal phyla: Cnidaria, Placozoa, Porifera and Ctenophora. Front. Physiol.

Sigwart, J.D., Sutton, M.D., 2007. Deep molluscan phylogeny: synthesis of palaeontological and neontological data. Proc. Royal Soc. B: Biol. Sci. 274, 2413—2419.

Smith, E.F., Nelson, L.L., Strange, M.A., Eyster, A.E., Rowland, S.M., Schrag, D.P., Macdonald, F.A., 2016. The end of the Ediacaran: two new exceptionally preserved body fossil assemblages from Mount Dunfee, Nevada, USA. Geology 44, 911—914.

Smith, C.L., Varoqueaux, F., Kittelmann, M., Azzam, R.N., et al., 2014. Novel cell types, neurosecretory cells, and body plan of the early-diverging metazoan *Trichoplax adhaerens*. Curr. Biol. 24, 1565—1572.

Sperling, E., Vinther, J., Pisani, D., Peterson, K., 2008. A placozoan affinity for *Dickinsonia* and the evolution of late Precambrian metazoan feeding modes. In: Cusack, M., Owen, A., Clark, N. (Eds.), Programme with Abstracts, vol. 52. Palaeontological Association Annual Meeting., Glasgow, UK, p. 81.

Sperling, E.A., Vinther, J., 2010. A placozoan affinity for *Dickinsonia* and the evolution of late Proterozoic metazoan feeding modes. Evol. Dev. 12, 201—209.

Srivastava, M., Simakov, O., Chapman, J., Fahey, B., Gauthier, M.E., Mitros, T., et al., 2010. The *Amphimedon queenslandica* genome and the evolution of animal complexity. Nature 14, 720—726.

Stafford, R., 2010. Constraints of biological neural networks and their consideration in AI applications. Adv. Artif. Intell. 2010. Article ID 845723.

Steele, R.E., David, C.N., Technau, U., 2011. A genomic view of 500 million years of cnidarian evolution. Trends Genet. 27, 7—13.

Takeda, N., Konc, Y., Artigas, G.Q., Lapébi, P., Barreau, C., Koizumi, O., et al., 2017. Identification of jellyfish neuropeptides that act directly as oocyte maturation inducing hormones. Development 145 (2), 156786.

Takeda, N., Konc, Y., Artigas, G.Q., Lapébi, P., Barreau, C., Koizumi, O., et al., 2018. Identification of jellyfish neuropeptides that act directly as oocyte maturation inducing hormones. Development 145 dev.156786.

Technau, U., Steele, R.E., 2011. Evolutionary crossroads in developmental biology: Cnidaria. Development 138, 1447—1458.

Thomson, D.W., Dinger, M.E., 2016. Endogenous microRNA sponges: evidence and controversy. Nat. Rev. Genet. 17, 272—283.

Valentine, J.W., Jablonski, D., Erwin, D.H., 1999. Fossils, molecules and embryos: new perspectives on the Cambrian explosion. Development 126, 851–859.

Watanabe, H., Fujisawa, T., Holstein, T.W., 2009. Cnidarians and the evolutionary origin of the nervous system. Dev. Growth Differ. 51, 167–183.

Whalan, S., Webster, N.S., Negri, A.P., 2012. Crustose coralline algae and a cnidarian neuropeptide trigger larval settlement in two coral reef sponges. PLoS One 7 (2012), e30386.

Wörheide, G., Dohrmann, M., Erpenbeck, D., Larroux, C., Maldonado, M., Voigtet, O., et al., 2012. Deep phylogeny and evolution of sponges (Phylum Porifera). In: Becerro, M.A., et al. (Eds.), Advances in Marine Biology 61. Elsevier Ltd. Academic Press, Amsterdam, pp. 1–78.

Wray, G.A., 2015. Molecular clocks and the early evolution of metazoan nervous systems. Philos. Trans. R. Soc. Lond. B Biol. Sci. 370, 20150046.

Xiao, S., 2013. Written in stone: the fossil record of early eukaryotes. In: Trueba, G., Montúfar, C. (Eds.), Evolution from the Galapagos - Social and Ecological Interactions in the Galapagos Islands. Springer, New York.

Xiao, S.H., Laflamme, M., 2009. On the eve of animal radiation: phylogeny, ecology and evolution of the Ediacara biota. Trends Ecol. Evol. 24, 31–40.

Yin, Z., Zhu, M., Davidson, E.H., Bottjer, D.J., Zhao, F., Tafforeau, P., 2015. Sponge grade body fossil with cellular resolution dating 60 Myr before the Cambrian. Proc. Natl. Acad. Sci. U.S.A. 112, E1453–E1460.

Young, G.A., Hagadorn, J.W., 2010. The fossil record of cnidarian medusae. Palaeoworld 19, 212–221.

Zakon, H.H., 2011. Evolution of sodium channels predates the origin of nervous systems in animals. Proc. Natl. Acad. Sci. U.S.A. 108, 9154–9159.

Zega, G., Pennati, R., Fanzago, A., de Bernardi, F., 2007. Serotonin involvement in the metamorphosis of the hydroid *Eudendrium racemosum*. Int. J. Dev. Biol. 51, 307–313.

Zhang, X., Shu, D., Han, J., Zhang, Z., Liu, J., Fu, D., 2014. Triggers for the Cambrian explosion: hypotheses and problems. Gondwana Res. 25, 896–909.

Zhang, S., Wang, X., Wang, H., Bjerrumb, C.J., Hammarlund, E.U., Costa, M.M., et al., 2015. Sufficient oxygen for animal respiration 1,400 million years ago. Proc. Natl. Acad. Sci. U.S.A. 113, 1731–1736.

Zhang, F., Romaniello, S.J., Algeo, T.J., Lau, K.V., Clapham, M.E., Richoz, S., et al., 2018. Multiple episodes of extensive marine anoxia linked to global warming and continental weathering following the latest Permian mass extinction. Sci. Adv. 4, e1602921.

Zumberge, J.A., Love, G.D., Cárdenas, P., Sperling, E.A., Gunasekera, S., Rohrssen, M., et al., 2018. Demosponge steroid biomarker 26-methylstigmastane provides evidence for Neoproterozoic animals. Nat. Ecol. Evol. 2, 1709–1714.

Chapter 3

Epigenetic requisites of the Cambrian explosion

Chapter outline

Advent of metazoans in the Cambrian explosion about 600 million years ago could not have been possible without evolving several basic mechanisms of epigenetic control of gene expression, such as epigenetic modifications of histones, DNA methylation, gene cooption, gene splicing, and evolution of epigenetic mechanisms of the immune defense, etc. These epigenetic processes are not spontaneous phenomena that occur randomly sensu genetic mutations,

Epigenetic Mechanisms of the Cambrian Explosion. https://doi.org/10.1016/B978-0-12-814311-7.00003-2

hence evident and self-explanatory, but highly determined in space and time, implying the use of information of some kind.

Commonly, the study of these epigenetic phenomena focuses on proximate causes and only exceptionally rarely follows the whole causal chain, which may reveal the "ultimate cause" or the ultimate source of the information we are looking for. An attempt is made in this chapter to shed some light on the causal chain and the source of information about the above epigenetic processes.

Gene recruitment

The use of an ancestral gene/gene regulatory network (GRN) for building a new structure and performing a new function in a new region of the body is known as gene/GRN recruitment, and the process is known as gene cooption. Gene cooption may induce sudden emergence of structural and functional novelties in animals. It is generally admitted that gene cooption/recruitment/deployment might have been a major evolutionary mechanism that contributed enormously to the evolution of novel biological functions (True and Carroll, 2002; Lenski et al., 2006).

Recruitment of a gene/GRN in strictly determined region of the body, organ, or tissue is not a random or accidental event, and hence information of some kind is invested for each particular recruitment event to occur in a particular region of the body. No model has ever been presented to explain what determines gene/GRN recruitment at a strictly determined region of the body.

It was believed that gene cooption is related with, or depends on, gene duplication (Lynch and Conery, 2000; Plachetzki and Oakley, 2007), which creates the possibility of neofunctionalization (acquisition of a new function by one copy of the gene), but this may not be true (or the whole truth) because cooption often involves not a single gene but simultaneously several genes. There is also empirical evidence that gene duplication is not necessarily related to gene cooption because generally it precedes gene duplication as is the case with crystallin genes where the appearance of the new function and cooption preceded the duplication (Piatigorsky and Wistow, 1991). A typical case of gene cooption without gene duplication is described by Harlin-Cognato et al. (2006), in which the enzyme astacin metalloprotease, without duplication of the gene, was used in the brood pouch of pregnant male pipefish, when the male brood pouch emerged as a novel body part, serving as a placenta providing nutrients to the embryo (Harlin-Cognato et al., 2006).

Evolution of lens crystallin genes

The vertebrate eye lens contains three crystallin forms, α-, β-, and γ-crystallins. Of these, α- and β-crystallins are related to small heat shock proteins, whose original function was to inhibit protein aggregation, by binding misfolded proteins, and to prevent cell apoptosis (Arrigo and Simon, 2010). β-Crystallin is the only one expressed in tissues outside the eye lens (Piatigorsky and Wistow, 1991), and α- and β-crystallins are encoded by the gene Cryab (Crystallin AB).

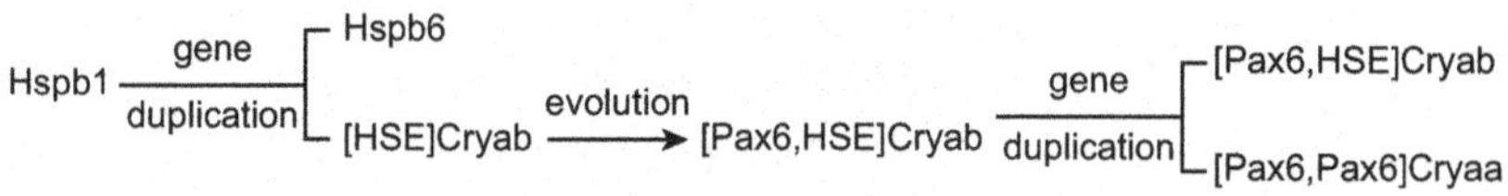

Hspb1 —— gene duplication ⌈ Hspb6
⌊ [HSE]Cryab ——evolution——→ [Pax6,HSE]Cryab —— gene duplication ⌈ [Pax6,HSE]Cryab
⌊ [Pax6,Pax6]Cryaa

FIGURE 3.1   The process of gene duplication originating from an ancestral *Hspb1* gene to produce a pair of *Cryab* and *Cryaa* genes. *From Cvekl, A., Zhao, Y., McGreal, R., Xie, Q., Gu, X., Zheng, D., 2017. Evolutionary origins of Pax6 control of crystallin genes. Genome Biol. Evol. 9, 2075—2092.*

How the heat shock binding element evolved into a binding site for Pax6 is not known, but it is believed that Hspb1, via gene duplication, evolved into Hspb6 and a [HSE]Cryab, with the last evolving into [Pax6, HSE]Cryab, which via another gene duplication led to emergence of genes *Cryab* and *Cryaa* (Cvekl et al., 2017), the first of whom HSE was converted into a Pax6 binding site (Fig. 3.1).

Neural involvement in recruitment of gene regulatory networks for eye development

Photoreceptor cells, simple light-sensitive cells capable of sensing light signals, may have evolved before the Cambrian explosion, but the spatial eye vision, eyes capable of forming visual images, emerged during the early Cambrian, 530—520 million years ago (Nilsson, 1996) (Fig. 3.2). Evolution of image-forming lenses was related to the cooption of cellular lenses in a number of independent cases from a few types of crystallin molecules (Wistow, 1993).

Experimental evidence shows that expression of *Pax6* in retina is indispensable for the initiation of lens fiber differentiation in mammalians (Reza and Yasuda, 2004a; Shaham et al., 2009; Iida et al., 2017) as well as for the development of retina itself (Reza and Yasuda, 2004b; Klimova and Kozmik, 2014). The lens-inducing role of *Pax6* is conserved across vertebrates and invertebrates as indicated by experimental evidence that murine *Pax6* induces formation of ectopic eyes in an invertebrate species, such as transgenic *Drosophila* (Graw, 1996).

During development, at the time the neural tube closes, it forms bilaterally two optic vesicles, the most anterior part of the forebrain, which narrows posteriorly to gradually develop into the optic nerve and grows anteriorly to contact the head ectoderm, inducing the latter to develop into the lens placode, whose invagination forms the lens pit. Then, signals from forebrain epithelium/optic vesicle, the precursor of the retina, induce transformation of the lens placode into lens (Furuta et al., 1998; Behesti et al., 2009). Besides, the adjacent brain is a substitute for the eyecup formed via invagination of the optic vesicle (Goss, 1969). The optic vesicle/retina releases diffusible crystallin-inducing signals, thus inducing crystallin synthesis (Cannata et al., 2008).

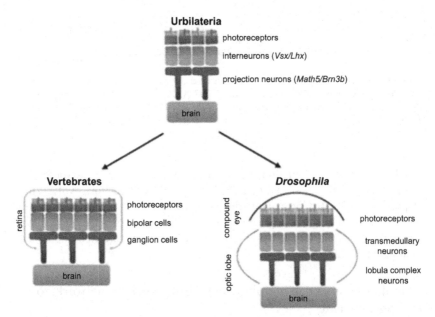

FIGURE 3.2   Model for the origin of the vertebrate and fly visual systems from a shared ancestral eye that already contained photoreceptors and their target neurons. Urbilateria contained ciliary (blue) (light gray in print version) and rhabdomeric (pink) (dark gray in print version) photoreceptors that targeted Vsx- and Lhx-positive interneurons. These neurons synapsed onto Math5- and Brn3b-positive projection neurons that targeted the motor centers of the brain. In the evolutionary line leading to vertebrates, the rhabdomeric photoreceptors were lost and the ciliary photoreceptors and their targets developed together in the retina. In the line leading to arthropods, the ciliary photoreceptors were lost and the progenitors for the rhabdomeric photoreceptors and the deeper neuronal layers were separated during development. Consequently, photoreceptors are located in the compound eye, whereas their neuronal targets are found in the optic lobe. Modified, with permission, from Erclik T., Hartenstein V., Lipshitz H.D., McInnes R.R. Conserved role of the *Vsx* genes supports a monophyletic origin for bilaterian visual systems. Curr. Biol., 18 (2008), pp. 1278–1287. *From Erclik, T., Hartenstein, V., McInnes, R.R., Lipshitz, H.D., 2009. Eye evolution at high resolution: the neuron as a unit of homology. Dev. Biol. 332, 70–79.*

Chemical signals from the optic vesicle directing the whole process from the development of lens placode to the formation of the complete lens are elements of a large complex GRN, comprising more than 2500 genes, which is responsible for eye development (Halder et al., 1995; Gehring, 1996) in *Drosophila*. In vertebrates, despite the large number of genes involved in eye development, within the lens developmental GRN, *Pax6* expression in the optic vesicle, the anterior part of the forebrain, plays the central role as a master gene.

It is believed that the development of lens "recapitulates" its evolution in vertebrates (Wistow, 1993). If true, this hypothesis implies that *Pax6* played a central role in the evolution of the vertebrate eye lens.

Induction of ectopic eyes (eyes developed in other regions of the body, such as antennae, wings, and legs) represents a form of gene cooption

(Nilsson, 2009). The experimental induction of ectopic eyes in *Drosophila* and even the ectopic expression of the mouse Sey (= *Pax6*) in *Drosophila* lead to ectopic development of eyes in the fly (Halder et al., 1995). Such cases of ectopic development of eyes imply incorporation of the gene(s) into a local GRN, enabling generation of a structure outside its natural region.

Among other genes involved in the coordinated process of lens formation are *Shh, Wnt, Bmp, Lhx Pax2, Vax1,* etc., with *Shh* in the midline restricting *Pax6* expression in dorsal regions where the optic cups develop (McLennan, 2008; Krishnan and Rohner, 2017) (Fig. 3.3).

Among the most important factors in lens development is another neural signal, *Lhx2,* released by the optic neuroepithelium that links several transcription factors and extracellular signals into a common genetic network that determines the transformation of the optic vesicle into optic cup (Yun et al., 2009) (Fig. 3.4).

Lhx2, released from the neural retina is required for optic vesicle patterning and lens formation (Yun et al., 2009), and deletion of *Lhx2* in embryonic neuroretina prevents lens development (Thein et al., 2016). These neuroretinal diffusible signals recruit a lens-specific GRN comprising three transcription factors (*Pax6, c-Maf,* and *Prox1*) and several crystallins to which other proteins join stagewise (Cvekl et al., 2017).

The lens GRN in extant animals is neurally activated by signals from the optic vesicle/retina (Reza and Yasuda, 2004a, 2004b; Weaver and Hogan, 2001; Arresta et al., 2005; Cannata et al., 2008; Rétaux and Casane, D., 2013), and there is reason to assume that the neural mechanism of activation of the lens development GRN in extant animals was operational during the Cambrian for at least two reasons:

Firstly, because Cambrian clades had evolved eyes, no selective pressure would arise for inventing an alternative GRN and mechanism for lens development. Evolution is not that wasteful as to invent another mechanism of developing the same structure.

Secondly, as predicted by the developmental constraint hypothesis, at this early stage of the embryonic development, the divergence of the early expressed genes is very limited (Wittkopp, 2007; Artieri et al., 2009).

Neural involvement in recruitment of proteins for *Bothrops* snake venoms

The snake *Bothrops jararaca* is a South American venomous viper. Production of the venom in its venom gland is induced neurally by signals from the sympathetic nervous system, which releases epinephrine in the snake's venom gland and via α- and β-adrenoceptors stimulate activation of transcription factors and the synthesis of the venom proteins (Yamanouye et al., 1997; Luna et al., 2009, 2013). The venom secretory system, thus, is "neuronally controlled" (Junqueira-de-Azevedo et al., 2015), and at different (activated or quiescent) stages, the gland produces different types of toxins (Luna et al., 2013).

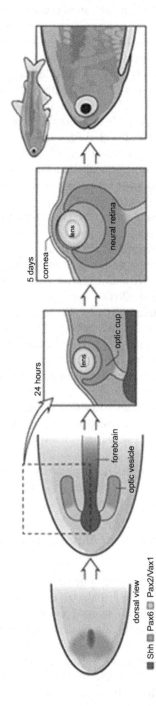

FIGURE 3.3 Eye development in surface and cavefish. The top row illustrates eye development in surface fish. Expression patterns of various key genes are color-coded. *From Krishnan, J., Rohner, N., 2017. Cavefish and the basis for eye loss. Philos. Trans. R. Soc. Lond. B Biol. Sci. 372: 20150487.*

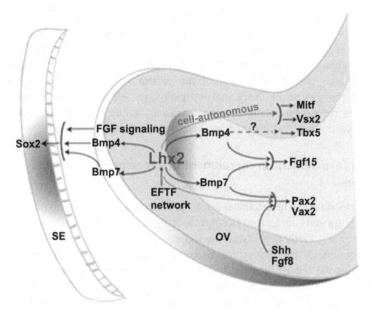

FIGURE 3.4   Model of *Lhx2* function during mouse early eye organogenesis. *Lhx2*, under the control of the EFTF (eye field transcription factors) network, links lens specification and optic vesicle patterning through the regulation of BMP signaling (*black arrows*). *Lhx2* also promotes optic vesicle patterning by cell autonomous mechanisms (*red arrows*). Why Bmp4 fails to upregulate Tbx5 expression is not resolved (*dashed line*). The timing of action and influence of *Lhx2* on several pathways suggest that it acts to coordinate the multiple patterning events necessary for optic cup formation. *From Yun S., Saijoh Y., Hirokawa K.E., Kopinke D., Murtaugh L.C., Monuki, E.S. et al., 2009. Lhx2 links the intrinsic and extrinsic factors that control optic cup formation. Dev. Camb. Engl. 136, 3895–3906.*

Administration of isopreterenol, a synthetic β-adrenergic receptor agonist, changes the contents of the snake venom and venom gland (Nunes-Burgos et al., 1993), whereas administration of the alkaloid reserpine, an antagonist of sympathetic activity, blocks snake venom synthesis (Yamanouye et al., 1997). This and the fact that the venom production begins with signals from the nervous system suggest a role of the nervous system in the recruitment of genes for the venom in the venom gland.

From an evolutionary viewpoint, the fact that the mechanism inducing synthesis of the venom, a complex cocktail of toxins varies widely among different species of the *Bothrops* species (Cardoso et al., 2010), indicates that in the course of evolution the nervous system of different species adjusted the venom contents to species-specific needs. The hypothesis that the present neural mechanism inducing venom production in extant snake species is the same with the one that determined the ancestral cooption of genes responsible for venom proteins is the most parsimonious hypothesis on the evolution of the venom. Evolutionarily, it would make no sense that, after evolving a mechanism of production and secretion of venoms, these snakes, under no specific selective pressure, would bother to "invent" another developmental mechanism of venom production during the individual development. The lack of evidence on any relevant changes in genes involved in the evolution of the venom production also suggests that the evolution of the venom was induced by neuronally induced changes in the developmental pathways. All the above strongly suggests that the neural mechanism that produces venom in extant species is the same with the one that coopted venom genes in ancestral species.

Neural involvement in recruitment of sex pheromones in salamanders

Both vertebrates and invertebrates release chemoattractants known as pheromones (from the ancient Greek *phero* (φέρω) "to bring" and hormone, from *orme* (ὁρμή) "stimulation"). Plethodontid salamanders, comprising about 300 species (Wilburn and Swanson, 2016), offer typical examples of the role of male pheromones in stimulating the female receptivity and courtship behavior. The pheromone, consisting of a number of different proteins, is secreted by male skin glands and the mental gland. The pheromone mixture is delivered intradermally by large premaxillary teeth scratching the female skin. In the plethodontid species, *Plethodon shermani* (formerly *Plethodon jordani*), the pheromone is delivered to the female by "slapping" across her nares a male structure that develops from the mental gland (Wilburn and Swanson, 2016) (Fig. 3.5).

The main components of plethodontid salamander pheromones are SPF (sodefrin precursor-like factor) believed to have evolved more than 100 million years ago and PMF (plethodon modulatory factor), between 50–100 million years ago (Palmer et al., 2007).

The volatile- and nonvolatile pheromones are received by the vomeronasal neurons, which convey the information for processing in the olfactory center (Wirsig-Wiechmann et al., 2006), resulting in stimulation of the female's courtship behavior. Also, experimental delivery of PRF (plethodontid receptivity factor) during the courtship increases the female receptivity (Rollmann et al., 1999).

SPFs are a family of proteins that evolved via gene duplications (Treer et al., 2018) in the last common ancestor (LCA) members of the Salamandridae family, about 100 million years ago, before the evolution of the sodefrin. Sodefrin, the first pheromone identified in vertebrates, is cleaved as a decapeptide from larger SPF precursors (Janssenswillen et al., 2015). About 60 million years ago in the LCA of Plethodontidae and Eurycea evolved PMF, a pheromone multiisoform. Later, about 27 million years ago, plethodontid salamanders evolved the new pheromone, PRF. Cooption of pheromone genes occurred independently many times in plethodontids and salamandrids, and it is believed to have played an important role in salamander speciation (Doty et al., 2016).

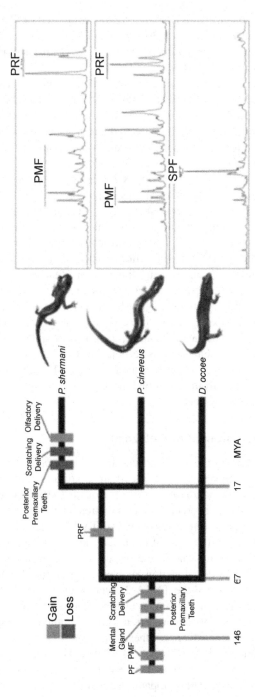

FIGURE 3.5 A reduced phylogenetic tree emphasizing species for which proteomic characterizations have been performed (*P. shermani*, *P. cinereus*, and *D. ocoee Plethodon shermani*, *Plethodon cinereus* and *Desmognathus ocoee*). Approximate times in millions of years (MYA) for gain and loss of key morphological, behavioral, or genetic trait acquisitions are included, as well as representative reverse phase chromatograms for the pheromones of each species. While plethodon modulatory factor (PMF) is one of the oldest pheromone gene families, its expression varies dramatically between the three species, with *D. ocoee* expressing a single PMF isoform at extremely low abundance (<1% total pheromone, not visible by HPLC/high performance liquid chromatography), *P. cinereus* expressing ~4–6 isoforms, and *P. shermani* expressing >30 isoforms. *From Wilburn, D.B., Swanson, W.J., 2016. From molecules to mating: rapid evolution and biochemical studies of reproductive proteins. J. Proteomics 135, 12–25.*

The composition of the pheromone mixture varies in different species. About 85% of the pheromone mixture in *P. shermani* is composed of PRF that appears in 3 isoforms and PMF in about 30 isoforms (Wilburn et al., 2015) compared to 4–6 isoforms in *P. cinereus* and a single PMF isoform in *D. ocoee* (Wilburn and Swanson, 2016).

In response to seasonal and visual stimuli, the male salamander secretes pheromones for courtship behavior ritual and slaps them to the female nares where they bind to the type-2 receptors of sensory neurons of the vomeronasal epithelium. These neurons transmit the stimulus to the neurons of the olfactory bulb where it is processed for assessing the isoform diversity and accordingly determine the intensity of the female courtship behavior.

The energetically expensive behavior in extant groups starts with the neurally determined secretion of the three groups of pheromones (SPF, PMF, and PRF) in the mental glands during courtship season. Curiously enough, the expression of the PRF gene in the mental gland of males occurred only in the eastern Plethodon species (Palmer et al., 2007). This, considered in the context of the neural control of the synthesis and secretion of pheromones and the fact that secretion of pheromones is triggered by innervation (Teal et al., 1989, 1999; Christensen et al., 1991), strongly suggests a crucial role of the nervous system in the ancestral cooption of the PRF gene beginning from ∼27 million ago. For if the cooption and expression of these pheromone genes in extant plethodontid salamanders is under neural control, there is no visible reason to doubt that the same neural mechanism was responsible for the evolution of cooption of pheromone genes in this group; there is no visible reason why ancestral species would switch from an original mechanism of cooption to the neural mechanism in the extant species.

Other cases of gene recruitment in metazoans

1. Recently is reported that in the small nematode *Caenorhabditis elegans*, the neurotransmitter octopamine, under certain conditions, undergoes succinylation leading to formation of osas#9 compounds (Artyukhin et al., 2013). Evolving initially as a process of elimination of waste products, later it was coopted for a new function. By binding TYRA-2 receptor of the nociceptive ASH neurons in the head of the *C. elegans*, it was specialized in detection of aversive stimuli and in the respective avoidance response (Hilliard et al., 2005). The new function evolved by coopting the octopamine degradation pathway and the TYRA-2 receptor in the ASH neuron of the worm, eliciting avoidance behavior even in starving worms, when the food patch is depleted. This is considered as an example of "cooption of intraorganismal neurotransmitter synthesis and pathways involved in interorganismal communication" (Chute et al., 2018). The binding of the deterrent pheromone signal osas#9 to the receptor of the neuron ASH induces avoidance behavior. The octopamine ascarosides, thus, connect intraorganismal neurotransmitter-based signaling pathways with interorganismal communication by creating a deterrent pheromone signal

(osas#9) from succinylated octopamine and the peroxisomally produced ascaroside ascr#9. In conjunction with the origin of ascarosides as end products of peroxisomal β-oxidation, the implication of octopamine succinylation as part of a general pathway for biogenic amine deactivation suggests that ascaroside biosynthesis may have originated as a pathway for waste product elimination that has been coopted for signaling functions.

2. Obligatory parthenogenesis is a widespread way of reproduction among many metazoan taxa, but the facultative parthenogenesis is relatively rare and individuals who switch to it maintain the ability of sexual reproduction despite the twofold cost of the sexual reproduction. It is observed that the facultative parthenogenesis appears in the absence of males, and because the switch to it is related to perception of the absence of males in the environment, the cascade of signals that determine morphophysiological changes related with the parthenogenetic reproduction has to begin in the nervous system. Examples of the facultative parthenogenesis are described recently in mayflies (Funk et al., 2010).

3. The planktonic crustacean, *Daphnia magna*, produces sexually- or parthenogenetically reproducing generations in response to unfavorable (crowding, depletion of food resources, or shortening of the photoperiod) (Stelzer, 2008) or favorable environmental stimuli, respectively. The switch to each of the GRNs determining production of eggs programmed for sexual and parthenogenetic reproduction is neurally determined: the environmental stimuli are perceived in the brain of *D. magna*, which reacts by activating a serotonergic pathway, starting a neurohormonal signal cascade that results in production by the mandibular organ of methyl farnesoate (MF), *Daphnia*'s juvenoid hormone, in the case of eggs programmed for sexual reproduction and suppression of MF secretion, and in the case of eggs that will produce the parthenogenetic generation (Wainwright et al., 1998). The switch to alternative reproductive modes is associated with differential expression of a number of genes, which is similar in various *Daphnia* species (Zhang et al., 2016). The synthesis and secretion of MF by the mandibular organ (Liu and Laufer, 1996) in *Daphnia* is negatively controlled by neuropeptides MOIH-1 (mandibular organ-inhibiting hormone 1) and MOIH-2 synthesized in, and secreted by, the secretory neurons of the X organ/sinus gland complex (Fig. 3.6).

4. The rotifer *Keratella cochlearis* responds to perceiving its small-sized predators *Asplanchna* spp. by elongating spines to make it more difficult for small predators to capture and swallow them, but it reduces spine length (and body size) upon perceiving the presence of larger predators, which prefer larger prey (Zhang et al., 2017; Sarma and Nandini, 2007) (Fig. 3.7). The response begins with perception, i.e., the signal pathway that determines development of the short or long spine in the individual *Keratella* or in its offspring is neurally triggered.

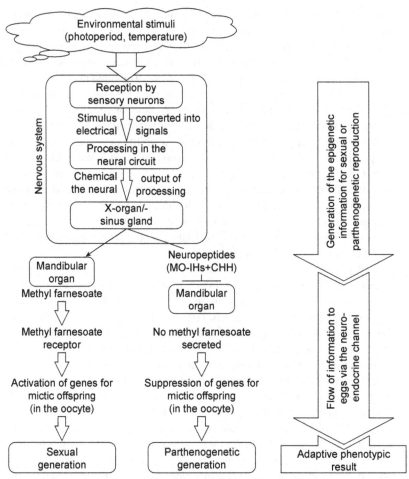

FIGURE 3.6 Illustration of the recruitment of genes in two different pathways leading to birth of asexual (parthenogenetic) and sexual generations in *Daphnia magna*. The point of divergence between the two pathways starts in the crustacean's brain with the failure of the X-organ/sinus gland to synthesize and secrete neuropeptides MO-IH1, MO-IH2, and CHH (left cascade) and recruitment of the methyl farnesoate (MF) in the case of sexual generation-producing Daphnias and recruitment of the previous genes and prevention of the synthesis of MF in the case of production of the asexual generation. *From Cabej, N.R., 2018. Epigenetic Principles of Evolution. Academic Press, London-San Diego- Cambridge, MA- Oxford, p. 355.*

In all the cases referred to in this section, the signals for inducing genes/GRNs and pathways responsible for respective characters originate in the nervous system. The most parsimonious and plausible hypothesis on the evolution of these cooptions is that the present-day neural mechanisms of the development were operational when their cooption first occurred, for evolution is parsimonious enough not to search for new "inventions" when no evolutionary pressure to do so exists.

(A)   **(B)**

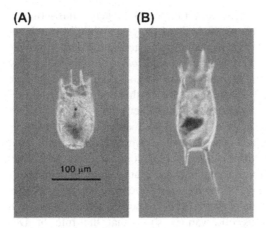

100 μm

FIGURE 3.7 *Asplanchna*-induced spine development in *Keratella slacki*. Live individuals photographed by the author at the same magnification. (A) Noninduced, basic morph with very short spines. (B) *Asplanchna girodi*—induced morph with pronounced spine development. *From Gilbert, J.J., 2016. Non-genetic polymorphisms in rotifers: environmental and endogenous controls, development, and features for predictable or unpredictable environments. Biol. Rev. Camb. Philos. Soc. 792. 964–992.*

Neural induction of epigenetic marks

Epigenetic marks, DNA methylation, and histone modifications represent a new layer of regulation of gene expression by determining the epigenetic profile of cells involved in the development, evolution, and intragenerational/transgenerational inheritance, pathogenesis of diseases, etc. Mechanisms of the placement of epigenetic marks are not known.

DNA methylation

DNA methylation is the process of the addition of a methyl group in DNA or its transfer to genomic new sequences (Cui and Xu, 2018). It is observed in almost all organisms, from unicellular prokaryotes to *Homo sapiens*. It is one of the most important epigenetic modifications involved not only in gene imprinting, X chromosome inactivation, modification of histones, and a number of other cellular processes but also in individual development and the maintenance of homeostasis and carcinogenesis.

DNA methylation is observed across eukaryotic plant and animal taxa, with rare exceptions, such as the nematodes *C. elegans* and *Pristionchus pacificus* and insects such as the mosquito *Anopheles gambiae* and *Drosophila melanogaster* (Yi and Goodisman, 2009). In metazoans, methylation affects almost exclusively C (cytosine), while in prokaryotes, A (adenine) is the base that is methylated most.

Methylation may occur in the gene body or in its promoter. When it occurs in the gene body, normally it is involved in expression of the gene, but when it takes place in the promoter, it represses the gene. In invertebrates, methylation predominantly affects the body of the gene, while in vertebrates, especially in mammals, it normally occurs in the gene promoter. However, recently, it is found that the invertebrate chordate *Ciona intestinalis* (the closest invertebrate relative of vertebrates) has a considerable number of methylated promoters in its genome, and, surprisingly, the methylated promoters in this chordate are associated with gene expression (Keller et al., 2016).

Somatic cells can transmit DNA methylation tags to their daughter cells and subsequent cell generations even when the causes that induced initially the appearance epigenetic changes are no longer present. DNA methylation occurs in almost all groups of animals and plants, from unicellulars to vertebrates. In plants and animals, the mechanism enabled gene imprinting. The global pattern of DNA methylation seems to be a vertebrate innovation (Albalat et al., 2012). Numerous observations show that the role of DNA methylation changed or expanded in the course of evolution and with the emergence of vertebrates (Cramer et al., 2017). This is related with the increased complexity of the vertebrate structural organization.

Commonly, the DNA methylation takes place in the sites where C is followed by a G (guanine) along the 5' → 3' direction. Such dinucleotides are known as CpG (5'—C—phosphate—G—3') or simply CG islands. DNA regions with high CpG frequency are known as CpG islands. Only about 25% of CpG islands are in gene bodies, whereas ∼50% of them reside in gene promoters.

Recently, it is reported on the occurrence of methylation in non-CpG sites (CpA, CpT, and CpC). The non-CpG methylation is more common in the human brain, where its frequency is inversely proportional to the level of the mRNA of the respective gene (Pinney, 2014). It occurs more often in the gene body and less in promoters and enhancers (Lister et al., 2009). The non-CG methylation is characteristic of embryonic stem cells; it is lost in the process of their differentiation but is restored when they are induced to dedifferentiate into pluripotent stem cells (Lister et al., 2009).

There seems to exist a positive correlation between the proportion of the methylated CpGs and the position of animals in the evolutionary ladder. In vertebrates, DNA methylation is the function of three DNA methyltransferases, Dnmt1, whose function is to maintain DNA methylation, and Dnmt3a and Dnmt3b that are responsible for adding new methyl groups onto DNA (Cui and Xu, 2018; Yap and Greenberg, 2018). Dnmt3 enzymes are present in all metazoans and plants (Zemach and Zilberman, 2010), while Dnmt2 proteins are present in plants, fungi, and protists, but they are also found in metazoans (Goll and Bestor, 2005). The function of Dnmt2 in metazoans is less known and controversial. This is the only Dnmt known in *Drosophila* and there is evidence that it is involved in a number of biological processes of the control of gene expression and in response to stressors in "Dnmt2-only" organisms, such as the fly (Vieira et al., 2018).

Neural control of DNA demethylation

During the last decade, it is reported that as a result of the neural activity, the protein GADD45a (growth arrest and DNA-damage-inducible protein) is involved in a mechanism of active demethylation based on the activity of TET protein family, (TET1−3) 5-methylcytosine hydroxylases, which oxidize 5mC into 5hmC (5-hydroxymethylcytosine) and cytidine deaminase APOBEC1 that converts C (cytosine) to U (uracil) (Guo et al., 2011) (Fig. 3.8). Conversion of 5mC into 5hmC is crucial element in the DNA demethylation process.

Sometimes, in the process of DNA replication, the methylated status in the newly synthesized strand may not occur, leading to passive demethylation (Wu and Zhang, 2010).

Under normal conditions, DNA methylation occurs in precisely determined DNA sites, but how DNA methylation systems recognize genes and gene sites they have to methylate is not known (Feng et al., 2010). A reasonable approach to this

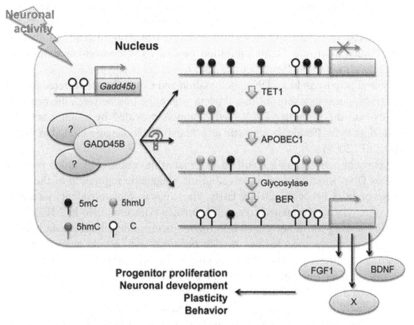

FIGURE 3.8 Neuronal activity−induced DNA demethylation in the adult brain. Neuronal activation in the dentate gyrus induces expression of Gadd45b, which plays an essential role in demethylation of Bdnf and Fgf1 promoters. This process also involves Tet1 and APOBEC1. Gadd45b-dependent DNA demethylation leads to expression of important genes, such as Bdnf and Fgf1, and can potentially have a broad impact in the mature nervous system. *APOBEC1*, DNA deamination enzyme; *BDNF*, brain-derived neurotrophic factor; *BER*, base excision repair mechanism. *From Guo, J.U., Su, Y., Zhong, C., Ming, G-l., Song, H., 2011. Emerging roles of TET proteins and 5-hydroxymethylcytosines in active DNA demethylation and beyond. Cell Cycle, 10:16, 2662-2668.*

question is to look for the proximate causes of methylation and follow upstream for a possible causal chain that might lead us to the source of that information.

DNA methylation is induced by various factors including neural signals. It is believed that the neuronal activity affects formation/repression of memory by inducing an intracellular cascade that activates the DNA methylation machinery by stimulating transcription of memory suppressors, while by repressing memory suppressors enables memory formation and synaptic plasticity (Heyward and Sweatt, 2015) (Fig. 3.9).

In water flea, *D. magna,* is observed a rare case of transgenerational effect of DNA methylation. After exposure to 5-azacytidine, an analog of cytidine, the crustacean experiences a decrease in global DNA methylation and this effect is transmitted to the next two unexposed generations (Vandegehuchte et al., 2010).

Learning leads to changes in DNA methylation patterns, which in turn participate in the memory consolidation process (Jarome and Lubin, 2014). Associative learning in rats causes gene-specific hypermethylation in hippocampus with the last helping to maintain memories (Miller et al., 2010). Neuronal activity and learning-dependent gene body methylation is essential for the development of reward-dependent learning and memory (Day et al., 2013). LTM (long-term memory) consolidation in *Aplysia* leads to specific gene methylation and Dnmt inhibition blocks LTM consolidation until its complete elimination (Pearce et al., 2017).

Neural activity induces DNA methylation and expression of secreted factors in mice hippocampal neurons in a process where the link between the neuronal activity and the promoter DNA methylation is provided by another activity-induced protein Gadd45b (growth arrest and DNA-damage-inducible, beta) (Ma et al., 2009).

Neuronal activity, as a result of external stimulation of neurons in vivo, induces DNA modifications involved in the epigenetic regulation of the adult mouse brain (Guo et al., 2011). Early life stress in female rodents increases DNA methylation compared to control females (Blaze and Roth, 2017).

Illuminating for the origin of information determining the methylation site in DNA are inherited changes in rats behavior induced in experiments of maternal LG (licking–grooming). Licking and grooming pups is an important postpartum maternal care behavior in these mammals. There are considerable differences in the LG behavior in rats: the frequency of LG in individual dams varies widely, from high to low LG frequency, although the total amount of time of dams in contact with their pups is similar in both high and low LG dams. Female rat puppies born to high LG mothers during the first week of life are reared by high LG mothers and also display high LG to their offspring. Similarly, rat puppies born to low LG mothers reared by low LG rat mothers show low LG behavior to their offspring. However, when female puppies of the low LG mothers are fostered to high LG mothers, they exhibit high LG to their offspring, and the reverse, female puppies born to high LG mothers are fostered to low LG mothers and show low LG behavior to their offspring.

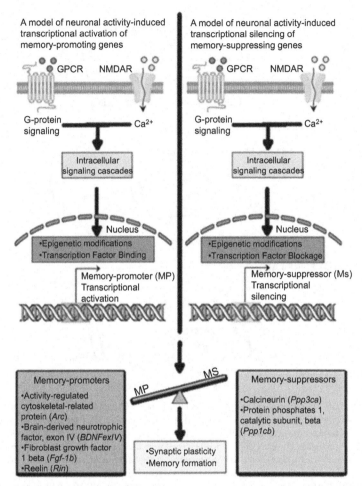

FIGURE 3.9 A model depicting the manner by which experience-dependent stimuli have been proposed to differentially regulate the expression of memory-promoter genes and memory-suppressor genes. Environmental stimuli, which consist primarily of associative learning tasks in animal models, evoke neurotransmitter-induced activation of specific postsynaptic receptors. Receptor activation stimulates specific intracellular signaling cascades that lead to distinct epigenetic patterns and transcriptional regulation at the gene regulatory domain of memory promoters and suppressors. The net increase in memory-promoter gene expression facilitates the establishment of synaptic plasticity and memory formation. List of memory-promoters: activity-regulated cytoskeletal-related protein (Arc), brain-derived neurotrophic factor, exon IV (BDNFexIV), Reelin (Rln), fibroblast growth factor-1 beta (Fgf-1b). List of memory-suppressors: calcineurin (Ppp3ca), protein phosphatase 1, catalytic subunit, beta (Ppp1cb). *GPCR*, G protein-coupled receptors, a transmembrane receptor; *MP*, memory promotion; *MS*, memory suppression; *NMDAR*, N-methyl-D-aspartate receptor, a neuronal ion channel receptor. *From Heyward, J.D., Sweatt, J.D., 2015. DNA methylation in memory formation: emerging insights. Neuroscientist 21, 475-489.*

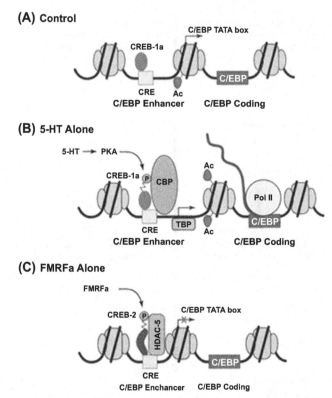

(A) Control

(B) 5-HT Alone

(C) FMRFa Alone

FIGURE 3.10 Diagram showing 5-HT (serotonin) and FMRFamide bidirectionally regulates histone acetylation(A) at the basal level, CREB1a resides on the C/EBP promoter; some lysine residues of histones are acetylated. (B) 5-HT, through PKA, phosphorylates CREB1 that binds to the C/EBP promoter. Phosphorylated CREB1 then forms a complex with CBP at the promoter. CBP then acetylates lysine residues of the histones. Acetylation modulates chromatin structure, enabling the transcription machinery to bind and induce gene expression. (C) FMRFamide activates CREB2, which displaces CREB1 from the C/EBP promoter. HDAC5 is then recruited to deacetylate histones. As a result, the gene is repressed. *From Guan, Z., Giustetto, M., Lomvardas, S., Kim, J.H., Miniaci, M.C, Schwartz, J.H. et al., 2002. Integration of long-term-memory-related synaptic plasticity involves bidirectional regulation of gene expression and chromatin structure. Cell. 111, 483–493.*

Thus, a maternal behavior experienced during the first week of life in rat puppies leads to acquisition and transgenerational inheritance of a new behavior without any changes in genes (Meaney and Szyf, 2005; Champagne, 2008). The transgenerational change is related to relevant epigenetic changes occurring in DNA of the glucocorticoid receptor (GR) gene of the rat hippocampus and preoptic area: perception of the LG tactile maternal stimulation leads to release in puppies of serotonin, which by binding its receptor in the hippocampal neurons activates an intracellular transduction pathway resulting in synthesis of

NGFI-A (nerve growth factor inducible A gene), which by demethylating the GR gene stimulates its acetylation and increased expression of GR.

Evolution of DNA methylation

Sponges possess a methylation machinery in the form of a complex macro-molecular aggregate composed of MBD2/3 (methylcytosine-binding) protein and NuRD (nucleosome remodeling deacetylase). In vertebrates, the duplication of MBD2/3 led to formation of a family of homologous proteins (MBD2, MBD3, MBD4, and MBD1+ MeCP2) and it is believed that the common ancestor of metazoans was in possession of the methylation-selective MBD2—NuRD complex (Cramer et al., 2017).

A shift in the patterns of methylation seems to have taken place with transition from invertebrates to vertebrates; while the latter display a global, genome-wide DNA methylation pattern characterized by intense methylation in CG sites, except for those in CpG islands, the invertebrates, along the plants and fungi, show a "mosaic" methylation with interspersed methylated and unmethylated domains (Feng et al., 2010).

The capability of using DNA methylation in gene regulation increased with the increase of the structural complexity of animals. As a general trend, the DNA methylation increased with the evolution of structural complexity. However, some taxa, such as the yeasts *Schizosaccharomyces pombe* and *Saccharomyces cerevisiae*, lack the methylation machinery, while other taxa such as *C. elegans* (Feng et al., 2010) and tunicate *Oikopleura dioica* have virtually lost the machinery and DNA methylation (Jeltsch, 2010).

Histone modification

Histones are chromatin proteins in the eumetazoan cell nucleus. The basic structure of the chromatin is the nucleosome, a core histone octamer consisting of two copies of each of the four histones (H2A, H2B, H3, and H4) around which a 147 bp segment of DNA wraps about 1.7 turns. Sequential nucleosomes are linked to each other through short "naked" DNA segments of about 80 base pairs long. The chromatin structure is generally repressive of gene expression chiefly because of the restrictive nucleosomal nature (LaVoie, 2005) and the packaging of the DNA into nucleosomes is partly responsible for its restrictive structure (Fig. 3.11). The nucleosomal core histones have protruded amino-terminal tails, which may posttranslationally undergo epigenetic changes (acetylation, methylation, phosphorypation, ubiquination, etc.) via enzymes such as histone acetyltransferases (HATs) for acetylation, histone methyltransferases for methylation, histone deacetylases (HDACs) for deacetylation (Fig. 3.12), and so on.

- H3
- H4
- H2A
- H2B
- H1
- DNA

FIGURE 3.11 Nucleosome structure and histone tail modifications. (A) The arrangement of the eight histone proteins in the nucleosome is shown schematically. One hundred and forty-seven base pairs of DNA are wrapped around the histone core. Histone H1 seals the nucleosome separating each nucleosome unit from each other. *From Hamon, M.A., Cossart, P., 2008. Histone modifications and chromatin remodeling during bacterial infections. Cell Host Microbe 4, P100-P109.*

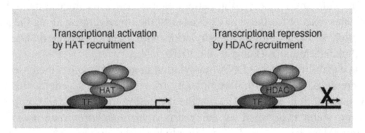

Transcriptional activation by HAT recruitment

Transcriptional repression by HDAC recruitment

FIGURE 3.12 Gene transcription is regulated by histone acetyltransferases (HATs) and histone deacetylases (HDACs). HATs and HDACs are recruited to the DNA by transcription factors (TF) that recognize specific DNA sequences. HATs and HDACs are part of multiprotein complexes (associated proteins indicated in gray). Increased transcriptional activation is associated with recruitment of HATs such as transcriptional coactivating proteins p300 and CBP (CREB-binding protein) and usually involves acetylation of core histones. HDACs such as HDAC1 and HDAC2 not only decrease transcription by reducing acetylation caused by stimuli but also may repress/ silence a gene by preferential recruitment to the regulatory region of the gene. *From LaVoie, H.A. 2005. Epigenetic control of ovarian function: the emerging role of histone modifications. Molecular Cellular Endocrinology 243, 12–18.*

Two general relationships are observed in regard to the accessibility of the genome:

- A strong inverse correlation between the DNA methylation and chromatin accessibility (the increased methylation negatively associated with chromatin accessibility in >97% of cases) and
- The constitutive hypomethylation is associated with variable chromatin accessibility (Thurman et al., 2012).

For DNA segments to be transcribed, they have to be accessible to transcription factors. The accessibility of genes to transcription factors is determined by the state of the nucleosome. Of special importance in determining the accessibility of genes to transcription factors is their epigenetic state determined among other things by the epigenetic changes.

Neural regulation of histone modifications

From earlier evidence on the relationship between the nervous system and induction of epigenetic marks, it was concluded that the enzymatic machinery inducing epigenetic modifications operates under the control of a variety of neuronal stimuli, which link physiological variations to modulated chromatin remodeling and thereby controlled gene expression (Borrelli et al., 2008). The injury stress and environmental stimuli also influence histone modification and associating chromatin remodeling via the nervous system (Fig. 3.13).

The pituitary follicle-stimulating hormone (FSH) is secreted in response to GnRH (gonadotropin-releasing hormone) secreted by hypothalamic neurons. Via body fluids, it reaches the ovarian granulosa cells and binds its specific membrane receptor, FSHR, starting an intracellular transduction pathway that leads to expression of FSH-responsive genes by phosphorylating histone H3 and CREB (Salvador et al., 2001) (Fig. 3.14).

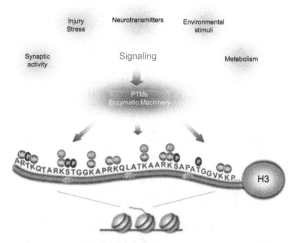

FIGURE 3.13 Multiple posttranslational modifications on histone tails. The H3 N-terminal tail, here presented as a paradigm of all histone tails, can undergo numerous modifications. Here, only phosphorylation, acetylation, and methylation are indicated. Methylation can be mono-, di-, or trimethyl. The enzymatic machinery that elicits these PTMs (posttranslational modifications) is believed to be under the physiological control of neuronal stimuli. Specific combinations of PTMs correspond to selective states of chromatin, either permissive or not for transcription, responsive to damage and stress, or modulated by physiological changes in cellular metabolism. *From Borrelli E., Nestler E.J., Allis C.D., Sassone-Corsi P., 2008. Decoding the epigenetic language of neuronal plasticity. Neuron 60, 961-974.*

Pulses of the neurotransmitter serotonin in *Aplysia* facilitate the LTM (long-term memory), whereas pulses of the neuropeptide FMRFamide cause its depression, but when a neuron receives input from serotonin, then the FMRFamide-induced depression is dominant. The reason is that the signals of serotonin and FMRFamide converge in the promoter C/EBP gene, where serotonin performs histone acetylation by activating CREB1, which recruits CBP (CREB-binding protein), while FMRFamide displaces CREB1 and recruits HDACS acetylation histones and blocks the facilitation (Guan et al., 2002) (Fig. 3.10).

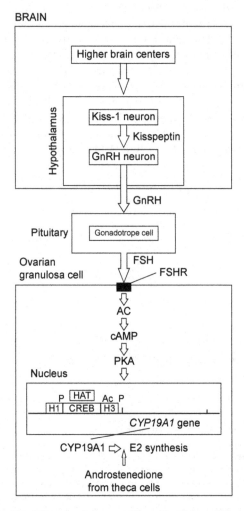

FIGURE 3.14 Simplified diagram of the neural control of histone H3 acetylation and phosphorylation, enabling induction of FSH-responsive genes and the synthesis of estradiol by granulosa cells surrounding the oocyte. Note that the epigenetic information necessary for H3 acetylation flows from hypothalamic neurons via FSH-secreting pituitary cells to ovarian granulosa cells where it induces FSH-responsive genes. *AC*, adenyl cyclase; *Ac*, acetylation; *cAMP*, cyclic AMP (3′-5′cyclic adenosine monophosphate); *CREB1*, cAMP response element binding protein one; *FSH*, follicle-stimulating hormone; *FSHR*, FSH receptor; *GnRH*, gonadotropic-releasing hormone; *P*, phosphorylation; *PKA*, cAMP-dependent kinase. *From Cabej, N.R., 2014. On the origin of information in epigenetic structures in metazoans. Med. Hypotheses 83, 378–386.*

Neuronal activity recruits HDAC2 (histone deacetylase 2) and histone—lysine N-methyltransferase (Suv39h1) by phosphorylating p66α via AMP-activated protein kinase, thus establishing repressive histone markers (Ding et al., 2017). Neuronal activity regulates gene transcription in synapses by inducing specific changes in nucleosomes or higher-order chromatin structure, thus making gene regulatory sequences accessible to DNA-binding proteins that have affinity for these sequences (West and Greenberg, 2011).

Neuronal activity in mice induces changes in chromatin accessibility of neurons, with ensuing dynamic changes in patterns of gene expression, although these changes are transient and reversible when compared with DNA methylation (Su et al., 2017) (Fig. 3.15).

The sensory activity in mice reduces expression of the histone H2BE in the chromatin of the olfactory neurons inducing transcription alterations in neurons of adult olfactory epithelium (Santoro and Dulac, 2012). In mice, learning (via classical conditioning) induces genome-wide increase in chromosome accessibility at 2365 regulatory (mostly promoters) regions (Koberstein et al., 2018).

Contextual fear memory, by activating NMDAR (N-methyl-D-aspartate) receptor and ERK (extracellular signal—regulated kinase), induces acetylation of histone in the area CA1 of the hippocampus (Levenson et al., 2004), and extinction of conditioned fear leads to increased acetylation of histone H4 around the BDNF P4 gene promoter (Bredy et al., 2007).

In response to neural activity, NuRD (nucleosome remodeling and deacetylase complex) induces deposition of H2A.Z variant at the promoters of a large number of genes and inactivates their expression in the brain (Yang et al., 2016; Su et al., 2017) (Fig. 3.16).

Learning and transient neuronal activation increases genome-wide chromatin accessibility in adult mammalian brains inducing changes in gene expression via an epigenetic mechanism, and the learning-regulated regions are negatively related to the development of ASD (autism spectrum disorder) that is characterized, among other things, with learning impairments (Su et al. (2017).

Stress pathway is a neurohormonal pathway proceeding mainly along the HPA (hypothalamic—pituitary—adrenal) axis that is characterized by changes

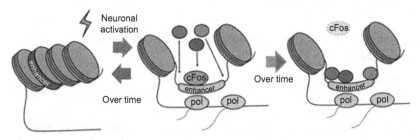

FIGURE 3.15   A working model for neuronal activity-induced changes in chromatin accessibility and dynamic changes in neuronal gene, where cFos is required for the initiation, but not the maintenance, of neuronal activity-induced chromatin opening. *From Su, Y., Shin, J., Zhong, C., Wang, S., Roychowdhury, P., Lim, J. et al., 2017. Neuronal activity modifies the chromatin accessibility landscape in the adult brain. Nat. Neurosci. 20, 476—483.*

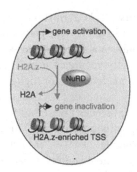

FIGURE 3.16 Model of the NuRD complex and H2A.Z chromatin remodeling link in the regulation of activity-dependent transcription and neural circuit assembly and function. *From Yang, Y., Yamada, T., Hill, K.K., Hemberg, M., Reddy, N.C., Cho, H.Y. et al., 2016. Chromatin remodeling inactivates activity genes and regulates neural coding. Science 353, 300–305.*

in the gene expression and behavior. The primary stress response involves behavioral changes and increased alertness, as well as physical and perceptive agility. This is followed by the secondary response that tempers the primary response and tends to reestablish the perturbed homeostasis. Under stress conditions, neurons induce epigenetic changes in chromatin (Rusconi and Battaglioli, 2018) (Fig. 3.17).

There is a critical period when juvenile male zebra finches can memorize the song of a "song tutor," but after that critical period, they cannot learn any more. To the contrary, juvenile males that have not heard a tutor during the critical period can learn from the tutor beyond the critical period. It was found that the loss of ability to learn after the critical period in the first group was result of epigenetic repression of the chromatin in birds' auditory forebrain (Kelly et al., 2018).

Recent studies on the network of histone modifications in sponge *Amphimedon queenslandica* have shown that they are pretty similar to higher animals, suggesting that this form of epigenetic regulation of gene expression existed at the advent of the multicellular animal life (Gaiti et al., 2017).

The above evidence is more than adequate to illustrate the instructive role of the nervous system in determining the precise location of epigenetic tags in DNA molecule and chromatin.

Epigenetic control of expression patterns and function of microRNAs

MicroRNAs (miRNAs) are a class of small (about 22–26 nucleotides long), single-stranded noncoding and highly conserved RNAs (Jeng et al., 2009) that posttranscriptionally block gene translation by binding complementary regions of target mRNAs (messenger RNAs) (Bethke et al., 2009; Sambandan et al., 2017) and are involved in the process of gene silencing. However, "under certain conditions, miRNAs can be also involved in gene transcription and

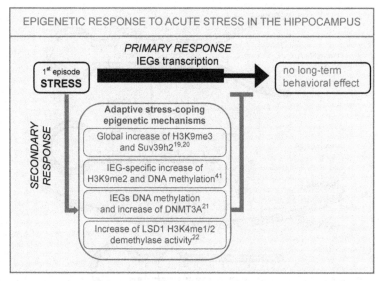

FIGURE 3.17 Epigenetic mechanisms of acute stress allostasis. Tolerable stress with a positive behavioral outcome elicits plasticity-gene transcription in the hippocampus instrumental to memorizing threat-related aspects of the negative experience with a protective valence. Meanwhile a set of epigenetic mechanisms buffer stress-induced transcription retaining its functional outcome within adaptive range. These mechanisms include global increase of H3K9me3 repressive histone methylation, increased levels of DNA methyltransferase Dnmt3a, as well as its association to the immediate early genes (IEGs) promoters and increased lysine-specific demethylase 1 (LSD1)-related repressive potential toward the same gene targets. This set of acute stress—induced epigenetic modifications contributes to counteract long-term behavioral effects on a cognitive or emotional point of view. *From Rusconi, F., Battaglioli, E., 2018. Acute stress-induced epigenetic modulations and their potential protective role toward depression. Front. Mol. Neurosci. 11, 184.*

translation" (O'Brien et al., 2018). They represent a new level of gene regulation. The first miRNA was discovered in 1993 (Lee et al., 1993).

miRNA genes are transcribed into primary miRNA (pri-miRNA), which is cleaved by a complex formed of the ribonuclease Drosha and Pasha that binds the complementary single-stranded fragments of the pri-miRNA to produce precursor miRNA (pre-miRNA), which is processed by Dicer and separated from the double strand by helicase enzymes to a smaller mature miRNA (Morlando et al., 2008) (Fig. 3.18).

The number of identified miRNAs is still growing. In humans, it increased from 2588 in 2008 (Gebremedhn et al., 2015) to 3707 in 2015 (Londin et al., 2015).

Evolution of miRNAs

The similarity of certain miRNA sequences among sea anemone, fly, and vertebrates suggests that the origin of miRNAs dates back to ~600 million years ago or to the LCA of animals (Technau, 2008).

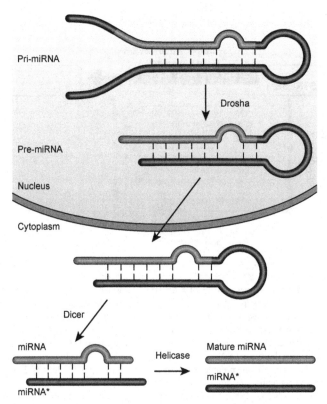

FIGURE 3.18 Diagrammatic representation of formation of a miRNA from a primary miRNA. *From Chen, K., Rajewsky, N., 2007. The evolution of gene regulation by transcription factors and microRNAs. Nature Reviews 8, 93–103.*

The oldest known miRNA, mir-100, and the related miR-125 and let-7 were initially active in secretory neurons around the mouth of bilaterians and cnidarians and only later in evolution their expression expanded into other tissues (Christodoulou et al., 2010).

The expression level of the earlier-evolved miRNAs is higher, more expanded, and generally more critical than those that evolved later. For instance, miR-1 and miR-208 are both involved in the heart development; the earlier-evolved miR-1 is expressed in the heart and muscles of vertebrates, whereas expression of miR-208 is restricted to only heart. Suppression of *miR-1* in mice is lethal, whereas miR-208 suppression, under normal conditions, results in normal development.

Three main mechanisms are thought to have been involved in the expansion of metazoan miRNAs (Hertel et al., 2006):

1 *De novo* synthesis based on the action of RNA polymerase II on hairpin RNA structures,

2 Tandem replication in existing miRNA clusters (in vertebrates and placentals),

3 Genome-wide duplications (in vertebrates and teleost ancestors).

It is suggested that evolution and expansion of miRNAs played a role in increasing the structural complexity, the advent of new Bauplaene, in shaping the macroevolutionary history of metazoans (Hertel et al., 2006; Peterson, 2009) and the accelerated increase of the transcribed noncoding RNA inventory (Heimberg et al., 2008). It seems that over time, the epigenetic mechanism of miRNA repression took over considerable part of the control of gene expression (Fig. 3.19), and the abrupt spikes in the miRNA family acquisition rate and the morphological change coincide in the Cambrian (Fig. 3.20).

Most metazoan transcription factor families (Fox, Sox, T-Box, etc.) were present in the last common ancestor (LCA) of eumetazoans and sponges and only a few evolved early in eumetazoan evolution and some even lost in sponges (Peterson and Sperling, 2007). By contrast, no miRNA family is known to have existed in the LCA of metazoans and appearance of miRNA families increased with the increased structural and functional complexity of metazoans what made biologists to believe that a causal relationship may be at the base of this correlation (Wheeler et al., 2009).

The curve in the figure shows two main spikes in the number of miRNA acquisitions, corresponding to the advent of the vertebrates and the mammals, whose appearance represents exponential increases in metazoan complexity.

It is estimated that miRNAs represent 1%−5% of animal genes (Stark et al., 2005). The number of miRNAs described is increasing at an accelerating pace (de Rie et al., 2017). *Monosiga brevicollis*, one of the unicellulars most closely related to metazoans, and the metazoan *Trichoplax* have no miRNAs (Grimson et al., 2008), whereas the evidence on the miRNAs in sponges is questioned (Thomson and Dinger, 2016). The number of miRNAs in the rest of animals with nervous system continued to increase: *A. queenslandica* has 8 miRNAs, *Nematostella vectensis* 40, the planarian *Schmidtea mediterranea* 61, *C. elegans* 154, *D. melanogaster* 147, *Mus musculus* 491, and humans 677 (Kosik, 2009).

More than 1800 of the identified miRNAs are involved in 50% of the transcriptome (Shomron et al., 2009). Be it simply correlational or reflection of a causal relationship, the above figure shows an interesting coincidence of the maximum of the morphological change (Cambrian explosion) with the highest level of miRNA acquisition.

It has been estimated that over 60% of human protein-coding genes (Friedman et al., 2009; Ebert et al., 2012) and 15% of *D. melanogaster* (Grün et al., 2005) are targets of miRNAs, suggesting that they might play a role in fine-tuning gene expression at the posttranscriptional level. mRNAs that are targets of miRNAs show less cross-species variability in gene expression than

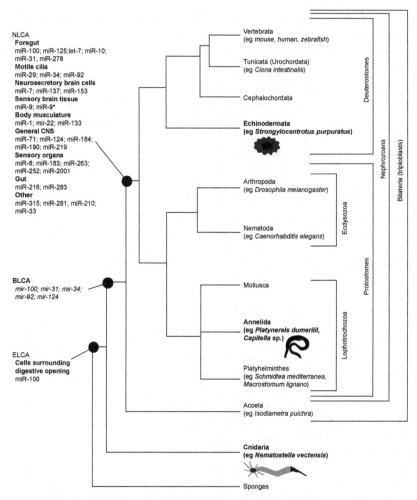

FIGURE 3.19   Phylogenetic relationships between major taxonomic phyla according to Egger et al. (2009) and reconstruction of ancestral tissue types based on conserved miRNA expression patterns. NLCA, BLCA, and ELCA: the Nephrozoan, Bilaterian, and Eumetazoan last common ancestor, respectively. The summary for the BLCA is preliminary owing to the absence of a sequenced acoel genome and miRNA expression data. Representatives of the taxa used in a study [Christodoulou et al. (2010)] are in bold. *From De Mulder, K, Berezikov, E., 2010. Tracing the evolution of tissue identity with microRNAs. Genome Biol. 11, 111.*

those not regulated by miRNAs and this may play a role in reducing noise in gene expression and establishing constraints in variability of gene expression (Cui et al., 2007).

Typically, the synthesis of miRNA starts with the synthesis of the primary transcript (pri-miRNA), which folds into a hairpin. Under the action of the

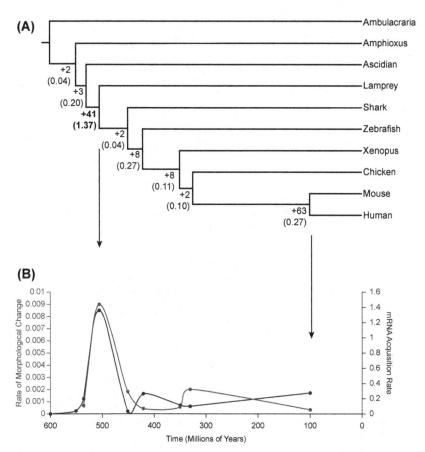

FIGURE 3.20 Evolutionary history of the 129 chordate-specific families of miRNAs found in eutherian mammals. (A) Cladogram derived from the history of miRNA family acquisition, with the number of new families indicated at the node and the rate of acquisition (number of new families per million years) shown parenthetically. Divergence times taken from estimates were derived from a molecular clock analysis (26) and the fossil record (44). (B) miRNA family acquisition rate (blue) plotted with rate of morphological change (2) (red) against absolute time. The spike for both miRNA acquisition and MCI (morphological complexity index) are both outliers (disproportionate higher) as compared with any other time in vertebrate history, as determined by a Dixon's D test. Points along the curves are tied to the nodes in A (two of which are indicated by arrows). *From Heimberg et al. (2008).*

enzyme Drosha, the transcript is transformed into a pre-miRNA in the hairpin form, which is released into cytoplasm where the enzyme Dicer trims off the loop to produce a miRNA duplex, with one of the strands (passenger strand) degraded and the remaining one (guide strand) incorporated into the RNA-induced silencing complex (RISC) with Argo (Argonaute) proteins as the active part (Desvignes et al., 2019) (Fig. 3.21).

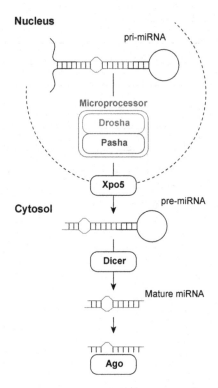

FIGURE 3.21 Schematic drawing of the canonical miRNA pathway in animals. Key proteins are indicated inside rectangles. *From Bråte, J., Neumann, R.S., Fromm, B., Haraldsen, A.A.B., Tarver, J.E., Suga, H. et al., 2018. Unicellular origin of the animal MicroRNA machinery. Curr. Biol. 28: 3288–3295.e5.*

Neural activity regulates miRNA expression

Spatial and temporal patterns of miRNA expressions are not random but strictly determined, indicating that information of some kind is involved in determining these patterns. To find the source of that information, there is no other way but to start with search for proximate factors or signals that induce miRNA expression. This search again leads us to the neural activity.

It is well-known that neuronal activity represses induction of miRNAs in neurons (Sambandan et al., 2017) and in other tissues and organs. In *Drosophila*, neuron activity regulates expression of miR-134 during the development of the neural circuitry (Fiore et al., 2009; Eacker et al., 2011). miR-219 and miR-132 in the mouse hypothalamic suprachiasmatic nucleus are induced in response to day light and regulate circadian timing in a process where miR-219 regulates circadian period length and miR-132 modulates light-induced clock resetting (Cheng et al., 2007). Neuronal activity regulates miRNA 132 expression in cultures of hippocampal neurons in mice (Wayman et al., 2008).

Mice respond to different levels of light with specific changes in the level of the miR-183/96/182 cluster, miR-204, and miR-211 in retinal neurons. The neurotransmitter photoreceptor neurons in this case are glutamate and it is observed that blocking glutamate, just like darkness, induces downregulation of these miRNAs; the reverse occurs during the light adaptation (Krol et al., 2010) (Fig. 3.22).

Mice exposed to artificial light at night show increased expression of miRNAs, miR-140-5p, 185-5p, 326-5p, and 328-5p, which by repressing expression of Rev-erba nuclear receptors in liver cells cause fat accumulation in the mouse liver (Borck et al., 2018).

Experimental depression in rats induces expression of miR-27a in the hippocampus and blood (Cui et al., 2018). Pharmacological (administration of pilocarpine) and behavioral (contextual fear conditioning and odor exposure) neuronal activation raised levels of miR-132 (pri-miR-132) and mature miRNA (miR-132) in the mouse brain (Nudelman et al, 2010).

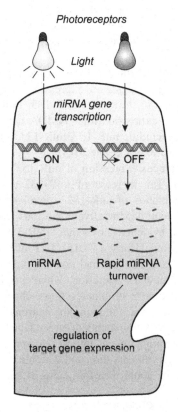

FIGURE 3.22 Simplified diagrammatic representation of the mechanism of light-induced miRNA expression in mouse photoreceptors. *From Krol, J. Busskamp, V., Markiewicz, I., Stadler, M.B., Ribi, S., Richter, J. et al., 2010. Characterizing light-regulated retinal MicroRNAs reveals rapid turnover as a common property of neuronal MicroRNAs. Cell 141, 618–631.*

Experimental post status epilepticus, induced chemically or electrically in rodents, showing similarity to human epileptogenesis, is characterized by differential expression of miRNAs: upregulated 274 miRNAs and downregulated 198 (Korotkov et al (2017). Pilocarpine-induced status epilepticus in rats leads to upregulation of 36 miRNAs and downregulation of 37 miRNAs (Araújo et al., 2016).

Neuronal activation induced experimentally by contextual fear conditioning, cocaine injection, and exposure to odorants, respectively, in the hippocampus, striatum, and olfactory bulb causes rapid and transient increases in expression of CREB-regulated miRNA-132 (Nudelman et al., 2010). Fear conditioning induces upregulation of 21 miRNAs (including miR-153, which is involved in both synaptic plasticity and LTM formation) in the hippocampus of the adult rats (Mathew et al., 2016). Activity-dependent long-term potentiation (LTP) and long-term depression are characterized by changes in temporal expression of 62 hippocampal miRNAs (Park and Tang, 2009). A study of 187 miRNAs in the hippocampus of contextually conditioned mice and in cultures of embryonic rat hippocampal neurons stimulated with NMDA or bicuculline showed that in both cases about half of the profiled miRNAs underwent similar changes (Kye et al., 2011).

Chemical induction of neuronal activity in the hippocampal neuron cultures was associated with changes in expression patterns of 51 miRNAs, including miR-134, miR-146, miR-181, miR-185, miR-191, and miR-200a, shows altered patterns of expression after NMDA receptor—dependent plasticity, and 31 miRNAs, including miR-107, miR-134, miR-470, and miR-546, were upregulated (van Spronsen et al., 2013).

Neuronal activity induces expression of miR-153 in rodent hippocampus (Mathew et al., 2016). The increase of miR-132 enhances cognition and memory, while downregulation of miR-132 is involved in the early stages of Alzheimer's and other neurodegenerative conditions (Salta and De Strooper, 2018), and experimental evidence shows that "miR-132 is part of an activity-dependent gene expression" (Hansen, 2015). LTP of synaptic strength in rat gyrus induces miR-132 and mir-212 and downregulates miR-219 expression levels in response to glutamate secretion, whereas inhibition of LTP with glutamate receptor antagonists eliminates these changes in miRNA levels (Wibrand et al., 2010). However, following the neuronal activity, initially most hippocampal murine miRNA transcripts decrease (Eacker et al., 2011).

A month after denervation, in rat muscle, downregulation of miR-1 and miR-133a occurs, but after reinnervation and 4 months after denervation, these same miRNAs increase by about threefold (Jeng et al., 2009).

miRNA regulation by neuropeptides

The neuropeptide Y (NPY) downregulates miR-375 in mice (Jeppsson et al., 2017), and the neurohypophyseal neuropeptide PRP-1 (proline-rich polypeptide-1) downregulates miR-302 (Galoian and Galoian, 2017).

The neuropeptide neurotensin, via its receptor NTR1, regulates differential expression of 38 miRNAs in human cancer colonocytes overexpressing the receptor (Bakirtzi et al., 2011) and via the transcription HIF-1α it upregulates miR-210 (Bakirtzi et al., 2016).

Diapause in insects is cerebrally regulated by the neurohormone DH (diapause hormone) and a few other neuropeptides. Based on the accumulating evidence that miRNAs are downregulated during diapause (Denlinger, 2002; Meuti et al., 2018), the fact that miRNAs are part of the "diapause toolkit" is concluded (Meuti et al., 2018).

Stress affects miRNA expression patterns

Stress conditions induce specific changes in expression of many miRNAs (Leung and Sharp, 2007, 2010), which act like mediators of stress response (Leung and Sharp, 2010). Expression of miR-10a and miR-21 in whole blood is regulated by acute psychological stress in humans (Beech et al., 2014).

Immediately after finishing exams, Japanese students had higher levels of miR-144/144* and miR-16 (Katsuura et al., 2012). Auditory fear learning-associated memory modulates miR-182 expression (Griggs et al., 2013) and auditory fear-learning suppresses miR-182 expression, by derepressing actin-regulating proteins (cortactin and Rac1), thus leading to consolidation of the long-term auditory memory (Murphy and Singewald, 2018).

Hormonal regulation of miRNAs

During the last decade, abundant evidence is accumulated on the hormonal control and regulation of miRNA expression. The evidence points to a correlation between expression of hormones and miRNAs. Certainly, correlational is not equivalent to causal, but solid experimental evidence now exists indicating that hormones control expression of specific miRNAs. So, e.g., in *Drosophila*, both in vivo at the onset of metamorphosis and in vitro in tissue cultures, ecdysone induces expression of two miRNAs, let-7 and miR-125 (Bashirullah et al., 2003). In the cotton earworm, injection of the neuropeptide DH, or one of its analogs, or ecdysone causes the termination of the diapause and this is associated by downregulation of miRNA-277-3. It is believed that termination of diapause induces downregulation of miR-277-3p whose targets are insulin-like peptides (LPs) secreted by brain neurons (Reynolds et al., 2018).

The neurohormone DAF-9 secreted by sensory neurons in *C. elegans* binds the nuclear receptor DAF-12 and determines the proper time of induction of miRNA let-7, which is responsible for larval development and longevity (Jia et al., 2002; Bashirullah et al., 2003; Bethke et al., 2009).

It is observed that the cerebrally regulated LH (luteinizing hormone) induces expression of miR-132, miR-212 (Fiedler et al., 2008), and miR-21 (Carletti et al., 2010) in murine granulosa cells and of miR-125a-3p in human granulosa cells (Grossman et al., 2015). FSH-induced progesterone secretion produces changes in expression of 31 miRNAs in granulosa cells (Yao et al., 2010). Administration of ACTH (adrenocorticotropic hormone) in mice induced specific changes in expression of four miRNAs (miR-96, miR-101a, miR-142-3p, and miR-433) (Riester et al., 2011). GC (glucocorticoid) treatment causes a global downregulation of miRNA expression in rat thymocytes during glucocorticoid-induced apoptosis of lymphocytes as a result of reduction of key microRNA processing enzymes (Smith et al., 2010), but treatment of bone cells with the glucocorticoid hormone, dexamethasone, and estrogen increases miR-17-92a expression in osteoblasts (Guo et al., 2013).

Pituitary and its target gland hormones, which are under ultimate neural control, are also involved in inducting miRNA expression. So, e.g., hormones of the HPA axis (ACTH and dexamethasone) change expression of circulating hsa-miR-27a in humans both in vivo and in vitro (Igaz, 2015). miRNA-125a and miRNA-455 are downregulated by ACTH in adrenals and ovarian granulosa (Hu et al., 2012a,b) and administration of hormones ACTH, 17α-estradiol, and dexamethasone in rats upregulates five and downregulates four miRNAs (Hu et al., 2013).

Folliculogenesis results from the coordinated activity of multiple genes in the ovarian follicle. The pituitary hormone FSH secreted in response to the hypothalamic neurohormone GnRH plays the determining role in differentiation and proliferation of granulosa cells by suppressing expression of three miRNAs, mir-143, let-7a, and mir-15b (Yao et al., 2009). It is also observed that the hypothalamically stimulated LH surge induces a dramatic upregulation of miR-132 and miR-212 in murine granulosa cells during ovulation (Fiedler and Christenson, 2007; Fiedler et al., 2008). The sex hormone, estradiol, induces 21 miRNAs and represses 7 others in the human breast cancer cells (Bhat-Nakshatri et al., 2009) and insulin downregulated expression of 39 distinct miRNAs in human skeletal muscle (Granjon et al., 2009).

Alternative splicing

The increased complexity of animal structure in the course of evolution is related to a larger amount and diversity of proteins. The relatively small number of proteins in early Cambrian metazoans triggered a selective pressure for a mechanism that could multiply the number and diversify the form of proteins. This led to evolution of the mechanism of alternative splicing, a process in which exons of a gene may be conserved or skipped during mRNA formation. Alternative splicing is an epigenetic mechanism (Holliday, 2002) of gene regulation and represents the main source of protein diversity in multicellular eukaryotes (Nilsen and Graveley, 2010; Lin et al., 2010).

Simply defined, alternative splicing consists in selective removal of introns and uniting remaining exons of a pre-mRNA (RNA precursor) into varied combinations, leading to production of numerous different proteins from a single gene. The splicing process is catalyzed by the spliceosome, a nuclear complex molecule comprising five small nuclear ribonucleoproteins (U1, U2, U4, U5, and U6) and many auxiliary proteins (Li et al., 2007), reaching a total of more than 150 proteins in the spliceosome complex. The conformation and composition of the nucleosome is highly dynamic (Will and Lührmann, 2011). The spliceosome recognizes specific pre-mRNA sequences, where it binds to excise introns (noncoding sequences) and ligate exons in various combinations, thus determining the sequence, nature, and conformation of resulting proteins.

Four major mechanisms of alternative splicing are known: exon skipping (SE), intron retention (RI), alternative 5' splice sites (A5'SS), and alternative 3' splice sites (A3'SS) splicing (McManus et al., 2014).

A correlation exists between two apparently unrelated sources of proteomic diversity, which is gene duplication and alternative splicing. Studies suggest that, after a duplication event, duplicates take over some amount of protein diversification via neofunctionalization, but in later stages again the role of alternative splicing increases, a phenomenon that is known as function sharing. It is suggested that the initial alternative splicing subfunctionalization may contribute to the neofunctionalization of the duplicate by increasing the number of protein isoforms (Su et al., 2006). As an illustration of the protein diversification potential of alternative splicing may be mentioned, Dscam (Down syndrome cell adhesion molecule). This is a family of immunoglobulin cell surface proteins necessary to form neural circuits. In *Drosophila*, it is estimated that Dscam1 gene may encode 38,016 different Dscam1 isoform proteins (Hattori et al., 2008).

In vertebrates, the brain is the organ where more alternatively spliced protein isoforms and splicing factors are expressed; some of them are expressed exclusively in the brain/nervous system, whereas others, while prevalent in the brain, are also found in other organs (Li et al., 2007). This may be related to the presence in this complex organ of thousands of highly specialized cell types that constantly undergo dynamic changes (Polydorides et al., 2000).

The number of genes in vertebrates and invertebrates is comparable. Given that the structural and functional complexity of vertebrates is incomparably greater, it is to be expected that alternative splicing must have a greater role to play in mammals and more complex in higher than in lower animal taxa. Indeed, there is evidence that confirms this prediction (Kim, 2007).

Alternative splicing is the major epigenetic mechanism of increasing the number of proteins that may be coded by single genes and it is estimated that it is involved in about 95% of human multiexon genes (Wang et al. (2008; Chen and Manley, 2009) and in 40% of *Drosophila* genes (Wang et al., 2015). In humans, more than 85% of the multiexon-analyzed genes contain at least one alternative splicing event (Lee and Rio, 2015; Mollet et al., 2010). The estimated number of full-length transcripts in humans is no less than 200,000 (Hu et al., 2015).

The best studied splicing factors are Nova1 and Nova2, which in mammalians are expressed only in the brain (Buckanovich et al., 1996; Irimia et al., 2011). Although Nova1 and Nova2 bind similar motifs, they differ as far as the location of their expression in the brain is concerned. While Nova2 is expressed throughout the brain, Nova1 expression is restricted to subcortical structures of the CNS

(Allen et al., 2010). Nova protein biochemical properties seem to be unchanged across bilaterians (Irimia et al., 2011). In the mouse brain alone, about 100 alternative exons regulated by Nova that binds to clusters of tetranucleotide YCAY motifs are identified (Jelen et al., 2007). A number of neurotransmitters and other neuroactive substances regulate production of splicing factors and their alternative splicing by modifying the intraneuronal Ca^{2+} (Xie, 2008). The nuclear RNA binding protein, SAM68, is another factor in the CNS that regulates splicing of a number of pre-mRNA targets, while SAM68 itself is regulated by the neural activity and CaMKIV (Iijima et al., 2011).

Neural regulation of alternative splicing

Under normal condition, alternative splicing determines that the right protein will be produced from a particular pre-mRNA. Production of the right protein rather than any randomly formed polypeptide clearly implies the use of specific information. To identify the source of the information used for the proper choice, first we have to determine the proximate cause of alternative splicing and follow the causal chain in the reverse course of events. In the following, adequate evidence on the determining role of the nervous system in alternative splicing in eumetazoans is presented.

A neural mechanism of accelerated production of splicing isoforms

The RNA polymerase elongation rate is the cause of the delayed transcript synthesis, but recently a new rapid neural "transcription-independent" mechanism of transcript synthesis is discovered. Some transcripts that retain select introns remain in the nucleus and under neural stimulation in due time undergo neural activity–dependent splicing, cytoplasmic export, and ribosome loading, thus sharply reducing the time of the synthesis of long transcripts from hours to minutes (Fig. 3.23). The neural mechanism of fast transcript synthesis involves NMDA receptor– and calmodulin-dependent kinase pathways and makes it possible to accomplish a normally hours-long process into a matter of minutes (Mauger et al., 2016).

The brain in vertebrates displays higher frequency of alternative pre-mRNA splicing compared to the rest of the organs and tissues (Su et al., 2018), and alternative splicing is commonly used in production of synaptic proteins.

Neuronal activity regulates alternative mRNA splicing Hermey et al. (2017). Different neural signals and different types of neurotransmitters lead to differential splicing variants according to the intracellular protein they select. So, e.g., it is observed that neurotransmitters dopamine and glutamate, determine formation of different ania-6 isoform proteins, a cyclin molecule encoded by the gene *ania-6* (Berke et al., 2001; Sgambato et al., 2003).

A proportion of fully transcribed RNAs retain introns (Braunschweig et al., 2014), and it is observed that in mouse neocortex neurons, these introns are excised as a result of neuronal activity (Mauger et al., 2016).

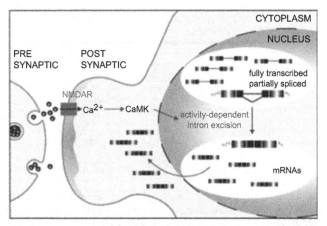

FIGURE 3.23   Transcripts containing activity-dependent introns are exported to the cytoplasm and associated with ribosomes. Working model. A subpopulation of polyA + transcripts retaining select introns is stored in the nucleus. Upon neuronal stimulation, these transcripts undergo rapid intron excision, and fully spliced mRNAs are exported to the cytoplasm, thus generating a readily available pool of mRNAs for translation. This process requires NMDA receptor and calmodulin-dependent kinase (CaMK) pathways. *From Mauger, O., Lemoine, F., Scheiffele, P., 2016. Targeted intron retention and excision for rapid gene regulation in response to neuronal activity. Neuron. 92, 1266–1278.*

The neurotransmitter glutamate increases the yields of splicing products (Reichert and Moore, 2000); it and other neuroactive substances induce changes in membrane potentials (Catterall, 2000), affect calcium channels, leading to elevation of Ca^{2+} levels, thus inducing specific changes in the pre-mRNA splice site selection by regulating chromatin modifiers (Sharma and Lou, 2011).

Neurexins are presynaptic membrane proteins encoded by genes *NRXN1*, *NRXN2*, and *NRXN3* in humans and *Nrxn1*, *Nrxn3*, and *Nrnx3* in mice, which produce long (α) and short (β) transcripts of the molecule. Alternative splicing of the neurexin pre-mRNAs can potentially produce more than 3000 alternatively spliced isoforms of the protein (Missler and Südhof, 1998; Ullrich et al., 1995) (Fig. 3.24).

In fear conditioning experiments, after repeated association of an electric shock with a particular sound rats responded to the sound alone and this was associated with a significant and transient (2 h) repression of SS#4 NRXN1/2/3α in the rat hippocampus, demonstrating the role of neurally determined changes in alternative splicing in the memory formation (Rozic et al., 2011). In the process of learning, the activity of hippocampal neurons modulates the splicing by accumulating a repressive histone marker, H3K9me3 (Ding et al., 2017). Psychological stress and fear in mice are associated not only with changes in gene expression patterns, but also with production of a novel alternatively spliced isoform termed Praja1a (Stork et al., 2001). In mice models of Rett syndrome, neuronal activity is

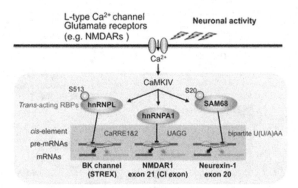

FIGURE 3.24 Model of the neuronal alternative splicing by neuronal activity and neuronal cell type—specific factors. (A) Depolarization or sustained neuronal activity activates CaMKIV signaling via NMDARs or L-type voltage-dependent calcium channels (L-VDCC). Splicing factors SAM68, hnRNPL, and hnRNPA1 are downstream of CaMKIV. These RBPs are thought be activated by CaMKIV and trigger a shift in alternative splicing of particular pre-mRNAs that contain their consensus recognition motifs. *CaMKIV*, calcium/calmodulin-dependent protein kinase type IV; *NMDAR*, N-methyl-D-aspartate receptor; *RBP*, RNA-binding protein. *From Iijima, T., Hidaka, C. and Iijima Y. 2016. Spatio-temporal regulations and functions of neuronal alternative RNA splicing in developing and adult brains. Neurosci. Res. 109, 1—8.*

found to cause aberrant mRNA splicing in the form of retained introns and skipped exons (Osenberg et al., 2018).

In one third of the cases of experimental ASD (autism spectrum disorder) in mice, occur changes in microexon splicing (microexon skipping), resulting from the neuronal activity that causes a decrease in the levels of the protein nSR100. Besides, it is suggested that even the nSR100 function depends on neuronal activity (Quesnel-Vallières et al., 2016) (Fig. 3.25).

It is interesting to point out that the vocal nuclei in the brains of birds capable of song learning (passerines, hummingbirds, and parrots) show differential splice forms of glutamate receptor superfamily genes, whereas doves that are a song nonlearner species do not (Wada et al., 2004; Marden, 2008).

Alternative splicing is also involved in performing behaviors in animals. For instance, in *Drosophila*, the courtship behavior is associated with activation of a circuit of about 1500 neurons distributed in clusters throughout the nervous system that express sex-specific splicing isoforms the fruM (fruitless) gene (Yu et al., 2010).

The mammalian gene *nogo* codes for three different proteins but denervation of mouse hind limb increases production of the nogo A mRNA and decreases nogo C mRNA. nogoA mRNA and nogoB mRNA are produced by alternative splicing, whereas nogoC mRNA by alternative promoter usage and investigators concluded that in both cases the splicing mechanisms in muscles "seem to be under neural control" (Magnusson et al., 2003).

Splicing also occurs in response to factors that block the neuronal activity (Mu et al., 2003) or stimulate excitatory activity (Li et al., 2007).

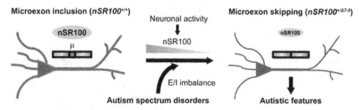

FIGURE 3.25 A model for activity-dependent nSR100 regulation and autism spectrum disorder (ASD). SR100/Srrm4 expression activates the splicing of a program of microexons in healthy neurons. Experimental stimulation of neurons to mimic increased neuronal activity observed in some ASD models and patient iPSC (induced pluripotent stem cell derived) neurons result in the rapid loss of expression of nSR100 protein and skipping of microexons (μ) that are also mis-regulated in autistic brains. Genetically reducing nSR100 levels reproduces the microexon splicing profile of activated wild-type neurons and autistic brains and causes autistic-like behavior as well as other ASD-like phenotypes in mice. *From Quesnel-Vallières, M., Dargaei, Z., Irimia, M., Woodin, M.A., Blencowe, B.J., Cordes, S.P., 2016. Misregulation of an activity-dependent splicing network as a common mechanism underlying autism spectrum disorders. Mol. Cell 64, 1023–1034.*

An anomalous alternative splicing characterized by intron retention and exon skipping occurs in response to neuronal stimulation (Osenberg et al. (2018).

Evolution of splicing

Yeast microscopic species *S. pombe* and *S. cerevisiae* diverged from their common ancestor 370 million years ago and from metazoans about one billion years ago, whereas mice and humans diverged from each other 75–130 million years ago. However, splicing factors are more similar between *S. pombe* and mammals than between the two yeast species. The alternative splicing phenomena observed in some unicellulars might represent exceptions or mis-splicing, so that it may be safer to say that alternative splicing is rare or absent in yeasts and unicellulars (Ast, 2004).

Focusing on the evolution of the Nova-regulated alternative splicing, Irimia et al. (2011) suggested that the first event in the rise of the Nova regulatory network was the evolution of the protein's ability to bind the YCAY (Y indicates a pyrimidine, U or C) repeats and perform splicing in the common ancestor of chordates and may be much earlier, in the common ancestor of bilaterians (Brooks et al., 2011). Then, in the tunicate ancestor of vertebrates, *Nova* expression was restricted to the brain, and in a third stage some new Nova-specific exons were added and regulation of Nova by preexisting exons coopted for Nova regulation (Irimia et al., 2011).

Immune defense: epigenetics of arms race with invading microorganisms

From the beginning, metazoans found themselves under threat of invasion by pathogenic microorganisms (viruses, bacteria and Protista). From their beginning,

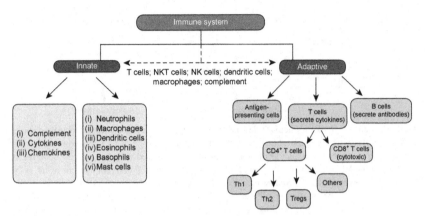

FIGURE 3.26 Overview of the immune system: innate and acquired immunity. An evolutionary bridge between both forms of immunity is observed because of the presence of γδ T cells, NKT cells, NK cells, dendritic cells, macrophages, and complement proteins. The innate immune responses include cells and soluble components that are nonspecific, fast-acting, and first responders in inflammation. In contrast, acquired immunity encompasses immune components that are more specific for targeted antigens and capable of forming immunological memory. *NK cells*, natural killer cells; *NKT cells*, natural killer T cells. *From Pandya, P.H., Murray, M.E., Pollok, K.E., Renbarger, J.L., 2016. The immune system in cancer pathogenesis: potential therapeutic approaches. J. Immunol. Res. Article ID 4273943.*

metazoans had to evolve mechanisms to target microbial invaders, without causing any harm to their own cells or their symbiont microorganisms, which implies mechanisms for distinguishing the "self" from "non-self" molecular patterns. The ability to distinguish between self and non-self at a molecular level is present in the oldest extant organisms, placozoans and sponges.

Immunity is function of the immune system, which comprises two subsystems, the innate (nonspecific) and the acquired (humoral/adaptive) immune systems (Fig. 3.26). The innate immune system is present in almost all animal taxa, invertebrates and vertebrates, whereas the acquired/adaptive system is characteristic of vertebrates alone. The immune system discriminates between "self" and "nonself," between organism's own molecular patterns and foreign molecular patterns as well as its own injuriously altered molecular patterns. The advent of the mechanism to discriminate self from nonself was the first step in mounting immune responses and a *necessary* condition for the evolution of the innate and acquired immune system. The multicellular life would be unsustainable without it; hence, the ability to distinguish between self and nonself, emerged simultaneously with the advent of the multicellular life or very soon thereafter.

Evolution of the mechanism to distinguish self from nonself marks the beginning of the innate immunity based on the presence from the beginning of the metazoan life of PRRs (pattern recognition receptors molecules) of the TLR (Toll-like receptor) type. TLRs are transmembrane receptors containing

LRRs (leucine-rich-receptors) capable of selectively binding several types of foreign lipoproteins, lipopolysaccharides, ssRNAs (single stranded RNA), dsRNAs (double stranded DNAs), etc.

Placozoans lack a complement system, but they have a component of the complement system, C1q-like factor (complement component 1q) and possess several proteins and putative LRRs that may perform TLR recognition functions (Kamm et al., 2019). This seems to be supported by the existence of downstream effectors of the canonical TLR pathway. One group of PRRs are the so-called scavenger receptors, mainly G protein–coupled receptors, that help in clearing the body from invading microbes and self-apoptotic cell debris in a process of endocytosis or non-opsonic phagocytosis (Kamm et al., 2019). The first line of defense in lower metazoans comprises also secreted proteins, such as interlectin proteins containing FReD (fibrinogen-related domains) proteins.

The TLR signaling is present in cnidarians and scavenger receptors in this clade acquired the function of host-symbiont recognition and regulation of symbiosis with the dinoflagellate *Symbiodinium minutum*, but it is possible that the symbiont itself blocks SR (scavenger ligand)-ligand-binding capabilities that, by preventing the cnidarian immune response, coopts the tolerogenic pathway (Neubaeur et al., 2017). On the other hand, there is evidence that lectins may provide a coat for bacteria and induce endosymbiosis in animals (Dinh et al., 2018; Neubauer et al., 2017).

Upon entering the cell cytoplasm, pathogens are recognized by NLRs (NOD-like receptors), which have domains for recognition and binding pathogens and their components. One of the most importants among them is the core apoptotic element, APAF-1 (apoptotic protease activating factor 1), of which *Trichoplax* harbors 12, and may be more, homologs. This is in contrast with cnidarians and other metazoans that possess only one APAF-1 homolog (Kamm et al., 2019). The presence of such a large number of APAF-1 homologs in this basal metazoan is probably related to the fact that it lacks other mechanisms that evolved later in the course of evolution.

Evolution of the innate immunity in invertebrates

Elements of the complement pathways appear first in cnidarians. The complement system in higher metazoans follows three pathways: the classical-, the lectin- and the alternative pathways. The complement effector molecules with the central protein C3, besides C2 and C4, appeared very early in metazoan evolution, in cnidarians, as a mechanism devoted chiefly to phagocytosis rather performing direct lysis of pathogens. Cnidarians also possess the key elements of the lectin pathway (Poole et al., 2016) (Fig. 3.27).

Expansion of the complement related genes occurred in a lineage specific manner, but expansion at the phylum and higher levels cannot be excluded.

Many invertebrates evolved MBLs (PRR mannose-binding lectins), ficolins, and the MASP (mannose-binding lectin associated serine proteases),

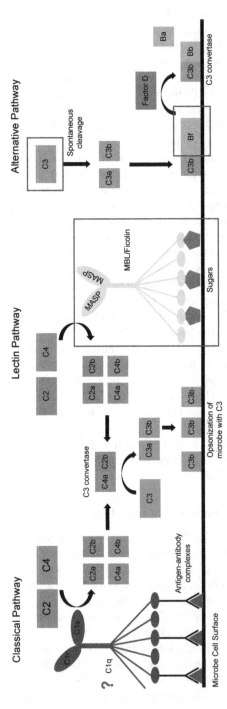

FIGURE 3.27 Diagrammatic representation of the complement system. In vertebrates, there are three activation mechanisms including the classical, lectin, and alternative pathways, which all converge at cleavage of the central protein C3. Invertebrates only possess components of the lectin and alternative pathways. Red boxes surround sequences that have been characterized in invertebrate genomes and transcriptomes. The question mark by C1q indicates that proteins with the C1q domain have been characterized in invertebrates, but a direct link to their role in the complement system has not been established. *From* Poole, A.Z., Kitchen, S.A., Weis, V.M., 2016. The Role of Complement in Cnidarian-Dinoflagellate Symbiosis and Immune Challenge in the Sea Anemone Aiptasia pallida. Front. Microbiol., 7 (519).

which is responsible for activating the lectin pathway (Iwaki et al., 2011), and complement factor B for the alternative pathway (Poole et al., 2016).

Important innate immune response proteins are discovered in particular sponges; MASPs and TEPs (thioester-containing proteins) are identified in marine sponges of the Class Homoscleromorpha (Riesgo et al., 2014). The sponge species *Oscarella carmela*, unlike other sponges possesses factors B and C3 (complement component 3).

Among cnidarians, a canonical Toll/TLR, is identified in the class Anthozoa (Miller et al., 2007). *Hydra vulgaris* has only one MASP gene but has neither C3 nor Factor B (Poole et al., 2016). It is suggested that the Toll/TLR pathway may predate the Porifera/Eumetazoa split (Miller et al. (2007).

In planarians, apparently the earliest bilaterian extant metazoans, took place an expansion of the complement system (alpha 2-macroglobulin and properdin), evolved antioxidant enzymes (catalase, glutathione peroxidase, etc.) and immunity-related proteins (placenta specific protein 8, mitogen-activated protein kinase (MAPK), interferon regulatory factor (IRF), etc.) (Gao et al., 2017). Planarians have a considerable amount of neoblasts, which are continuously differentiated into different (somatic and sexual) cell types necessary for the growth, regeneration, and for replacing missing cells. After a low dose X-ray irradiation only a few neoblasts survive in planarians, but soon thereafter, in the vicinity of the central nervous system and the nerve cord is activated a neoblast proliferation program to meet the requirements for specific differentiated cells. It is already demonstrated that neural signals are responsible for the process of proliferation and migration of neoblasts that restitute the eliminated neoblast population (Rossi et al., 2012).

In a recent study is reported about a form of "instructed immunity" in planarians infected with the Gram-positive bacterium *Staphilococcus aureus*. It shows the development in planarians of an immunological memory and heightened resistance to re-infection, believed to be related with a critical activity of neoblasts involving expression of the peptidoglycan Smed-PGRP-2 that induces expression of Smed-setd8-1-dependent expression of anti-bacterial genes during re-infection (Torre et al. (2017).

Studies on earthworms and leeches have shown that annelid immune defense is mainly based on the activity of various types of the immune cells/coelomocytes of the coelom rather than hemocytes (Salzet et al., 2006). In the earthworm, small coelomocytes (SCs) cause co-cultured foreign cells to lyse, implying that these annelids have a natural killer cell-like function similar to vertebrate NK (natural killer) cells. Larger coelomocytes (LCs) engulf and encapsulate lysed target cells. Thus, SCs are cytotoxic and LCs are phagocytic cells. In cytotoxic small coelomocytes, but not in large coelomocytes, are detected surface receptor molecules of the type identified in various vertebrate thymocytes (T cells), such as CD11A+, CD45RA+, CD54+, CD90, etc. A cytolytic factor isolated from the coelomic fluid of the earthworm *Eisenia fetida* has similar effect with the vertebrate TNF (tumor necrosis factor); it has two pattern recognition domains to bind to the pathogen cell wall components and then induce its lysis (Silerová et al., 2006; Salzet et al., 2006). In annelids for the first time appears an antigen-binding protein (Laulan et al., 1983).

Climbing on the evolutionary ladder the immunological arsenal of metazoans is enriched with new elements (Fig. 3.28).

Arthropods possess PRRs (pattern recognition receptors), which detect and bind invading microbes, activate effector proteins and stimulate their phagocytosis in hemolymph. However, the immune response in arthropods is still general and not specific to specific pathogens. TEPs (thioester-containing proteins) are the most important effectors in arthropods. Their binding to the pathogen helps phagocytes to identify, engulf, and cause lysis of the pathogen. In arthropods opsonin activity have lectins, FREPs (fibrinogen-related proteins), and the hypervariable IgG domain protein Dscam (Baxter et al., 2017).

Among insect TEPs (iTeps) (thioester-containing proteins), the protein fold of the malarial *A. gambiae*, TEP1, resembles that of the complement factor C3 (Baxter, 2007). The insect hemolymph contains many antimicrobial peptides predominantly secreted by the fat body, the equivalent of vertebrate liver. Six

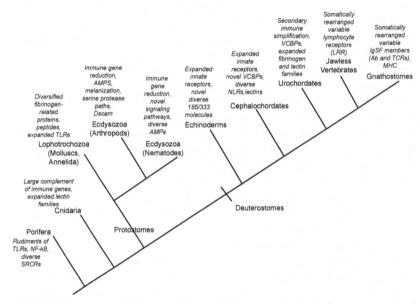

FIGURE 3.28 An overview of some of the novel features associated with immune responses of representatives of major animal lineages. *Ab*, antibodies; *AMP*, antimicrobial peptide; *Dscam*, Down syndrome cell adhesion protein; *IgSF*, immunoglobulin superfamily; *LRR*, leucine-rich repeat; *MHC*, major histocompatibility complex; *NLRs*, intracellular NOD-like receptors; *TCR*, T cell receptor; *TLR*, Toll-like receptor; *VCBPs*, variable region-containing chitin-binding proteins. *From Loker, E.S., 2012. Macroevolutionary immunology: a role for immunity in the diversification of animal life. Front. Immunol. 3, 25.*

of these peptides have been characterized in *Drosophila* and their genes are cloned (Meister et al., 1997). The protein Dscam (Down syndrome cell adhesion) detected in the *Drosophila* hemolymph belongs to the Ig (immunoglobulin) superfamily of receptors. Based on the structure of the *Dscam* gene, it is estimated that *Drosophila* immune-competent cells can potentially express about 18,000 different isoforms. It is demonstrated that many of these isoforms bind directly to bacteria suggesting that they serve as receptors or co-receptors during phagocytosis and are involved in the process of opsonization of pathogens. Inhibition of *Dscam* expression in hemocytes or blockage of Dscam interactions leads to impairment of phagocytosis (Watson et al., 2005).

Immune system in protochordates is function of morphologically and functionally different immune cell (coelomocyte) populations circulating in the coelomic fluid of the coelomic cavity, such as amebocytes, phagocytic cells, and filopodial cells that are responsible for clotting. Among humoral factors of the immune defense in echinoderms are lysins, agglutinins, lectins and other antimicrobial substances (Chiaramonte and Russo, 2015). In the cephalochordate amphioxus is identified an immunoglobulin superfamily (IgSF) gene homologous to *CD47* and a multigene family containing an Ig-like variable region (Pestarino et al., 2007), suggesting that protochordates may have evolved an incipient adaptive immunity.

Tunicates are the closest relatives of vertebrates (Delsuc et al., 2006) and in their hemolymph circulate phagocytes (immunocytes), which synthesize pattern-recognition receptors and are specialized in phagocytosis, encapsulation, and secretion of various lectins. The second type of the immune cells in their hemolymph are cytotoxic cells, specialized in cell-mediated cytotoxicity and release of cytokines, complement factors, and antimicrobial peptides. In protochordates, are identified both the alternative and lectin complement-activation pathways (Franchi and Ballarin, 2017).

The acquired (adaptive) immunity in vertebrates

The immune system in vertebrates is characterized by somatic mutation and long-term immunological memory with an increased and dominant role of the nervous system in the activation and development of the immune response.

Vertebrates inherited from their invertebrate ancestors and retain the mechanism of the innate immunity, a non-specific mechanism of immunity. However, the innate immune system, both its cellular and humoral arms, was inadequate and unable to protect Cambrian vertebrates from pathogens, as it is proven by the extant vertebrates. This acted as a selection pressure for evolving an additional mechanism of the immune response that is known as acquired or adaptive immune response, which is specific against any micro-organism, protein, or antigen. Thus, vertebrates invented a new mechanism capable of not only distinguishing the self from nonself, but of *recognizing* and *specifically* acting against any possible kinds of microbes, proteins, compound proteins (glycoproteins, lipoproteins, lectins, etc.).

Vertebrates are easily accessible and exposed to thousands of different microbes and microbe strains, but they are capable of producing specific antibodies against each of millions of different antigens that may gain access into their body. This form of the immune response is a strictly regulated and highly complex process requiring investment of huge amounts of information. Obviously, the vertebrate genomes comprising only 20–40,000 genes cannot be taken into account as source of that information.

In a unique evolutionary feat, about 500 million years ago, with the emergence of the jawed fish, the immune system acquired its adaptive arm, i.e., the ability to produce specific antibodies against practically any imaginable antigenic molecule or surface receptor. The number of the immune genes is relatively small and negligible when compared with the enormous number of *specific* antibodies vertebrates need to produce in response to various types of cells, cell components, and compound molecules they encounter. To meet this requirement vertebrates evolved the acquired immune system that uses two basic epigenetic mechanisms in lymphocytes of two types, B lymphocyte (B cells) and T lymphocytes (T cells) in lymphoid organs:

1. The recombination of the V(D)J segments (Variable gene segments, Diversity gene segments and Joining gene segments). Crucial element in the evolution of the acquired immunity played the proteins RAG1 and RAG 2 (recombination activating genes 1 and 2), which, as part of the V(D)J recombinase are involved in the generation of the huge number of diverse membrane receptors and antibodies by B cells. These proteins are believed to have evolved through invasion of a putative immunoglobulin-like gene by a retrotransposon (Janeway et al., 2001a; Janeway et al., 2001b). During the Cambrian, it is believed that "100 quadrillion possible different lymphocyte receptors arose seemingly overnight" (Laird et al., 2000).
2. Hypermutation of the hypervariable DNA regions of B cells that diversifies their membrane receptors by increasing up to one million times the average frequency of point gene mutations.

The rise of the myriad of specific antibodies and cell receptors in a blink of an eye during the Cambrian and the unimaginably high frequency of mutation rates have no genetic explanation and contradict the neo-Darwinian idea of the nature of molecular evolution and the frequency of gene mutations, while the fact that these changes are acquired during the life, but are not inherited, indicates that the somatic hypermutation in the process of development of the immunological competence is an epigenetic mechanism (Holliday, 2002).

Upon contact with a foreign antigen (pathogen, protein, compound protein, etc.) an antigen -presenting cell (a macrophage, a dendritic cell or B cell) binds the antigen and displays it in a recognizable form to a T cell. Then, a B cell that has surface receptors for the antigen takes up the antigen from the T cell via the receptor-mediated endocytosis and enters the state of immunoglobulin class switching, somatic hypermutation, and proliferation leading to formation of a clone of the B cells that produce exclusively a specific antibody against the specific antigen.

A different, but also sudden, acquired immune system evolved in the jawless fish, which lack B cell receptors (BCRs), T cell receptors (TCR), Major histocompatibility complex (MHC), and RAG1/2 proteins but use very different effectors (Flajnik and Kasahara, 2010) (Fig. 3.29).

Evolution of RAG1 and RAG2 genes

Discovery of transposons containing homologs of RAG1-RAG2 genes in the genome of the green sea urchin and starfish suggests that the direct ancestor of the vertebrate V(D)J recombinase was a hypothetical Transib VDJ that was recruited in the V(D)J machinery about 500 Ma, with the advent of jawed fish (Kapitonov and Koonin, 2015). A protoRAG containing both RAG1-like and RAG2-like displaying similarities with vertebrate RAGs dating back 550 million years ago is also identified in amphioxus (Carmona and Schatz, 2017) and biologists wonder why it was not coopted by other taxa in the course of evolution.

Neural control of the acquired immune response

During the two last decades of the last century, reports on the neural control of the immune response appeared. However, only in the last several years, the issue acquired the attention it deserves, and the results of the research have been encouraging and inspiring. It has long been known that the CNS controls immunity in vertebrates humorally, via HPA axis, what is a global control of the immune response. Now we know that an even bigger and determining role in the development of the immune response plays the autonomic nervous system, both the sympathetic and parasympathetic arms, involving, respectively, neurotransmitters epinephrine and acetylcholine. The autonomic nervous system thus emerges as a "messenger from the mind to the body for all organ systems, including the immune system" (Sanders, 2012).

The brain receives sensory information from the infection site on antigen recognition by immune cells in the site of infection via neural (sensory nerves of the sympathetic and parasympathetic nervous systems) and via nonneural (inflammatory substances, substances released by immune cells, and pathogens) pathways. Pathogens and immune cells in the infection site release cytokines, prostaglandins, and other substances that bind respective

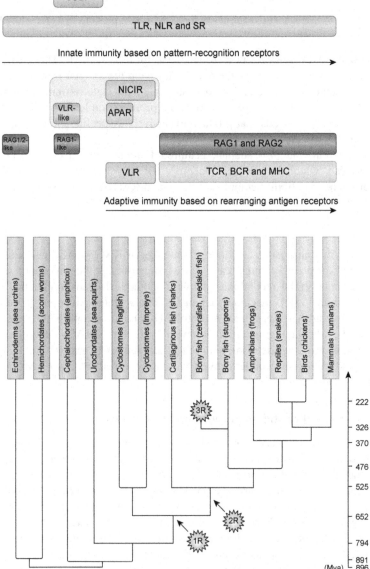

FIGURE 3.29 The stages in phylogeny of the emergence of immune molecules. Molecules restricted to jawed and jawless vertebrates are in nine right hand side vertical bars and horizontal bars above them. Molecules that emerged at the stage of invertebrates are in four left-hand side vertical bars and the uppermost full-size horizontal bars. Recombination-activating gene (RAG-like) genes are of viral or bacterial origin (from the transib transposon family) and are also present in the genomes of sea urchins and amphioxus. Agnathan paired receptors resembling antigen receptors (APAR) and novel immunoreceptor tyrosine-based activation motif-containing immunoglobulin superfamily receptor (NICIR), also known as T cell receptor (TCR), are agnathan immunoglobulin superfamily (IgSF) molecules that are thought to be related to the precursors of TCRs and B cell receptors (BCRs). The divergence time of animals is shown in Mya (million years ago). *Abbreviations: 1R*, first round of whole-genome duplication (WGD); *2R*, second WGD round; *3R*, lineage-specific WGD experienced by an ancestor of the majority of ray-finned fish 320 million years ago; *MHC*, major histocompatibility complex; *NLR*, Nod-like receptor; *SR*, scavenger receptor; *TLR*, Toll-like receptor; *VCBP*, V-region containing chitin-binding protein; *VLR*, variable lymphocyte receptor. *From Flajnik and Kasahara (2010).*

receptors in neurons of the autonomic nervous system, thus conveying information about the immune status to the brain. The information processed in the various brain areas is integrated in the hypothalamus, which activates the HPA axis and the autonomic nervous system (both the sympathetic and parasympathetic nervous systems). Neural signals stimulate egress from bone marrow of hematopoietic stem cells (Katayama et al., 2006) and accumulation of the immune cells around nerve fibers of the lymphoid organs (van de Pavert et al., 2009). Nerve fibers of the sympathetic nervous system secrete in lymph nodes norepinephrine, which binds to beta-2 adrenergic receptor (β2AR) of B cells, activating a transduction pathway and leading to a rise in the level of the IgG1 and CD86. Both signaling pathways converge to transcription of the mature immunoglobulin IgG1 (Figs. 3.30 and 3.31).

The result of the immune response is proliferation of a specific lymph B cell which enters a process of clonal selection, leading to formation of a clone of this cell type that is large enough to provide the organism with the huge amount of specific antibodies to nonself substance (usually cell surface receptors of the pathogen). The increase in IgG1 is not due to an increase in

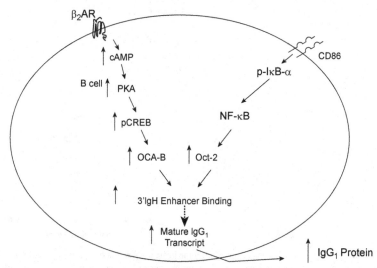

FIGURE 3.30 The beta-2 adrenergic receptor (β2AR) and CD86 signaling pathways converge to induce an increase in the level of IgG1 produced by an activated B cell. The β2AR engagement on an activated B cell activates cAMP/PKA/CREB, while CD86 engagement on an activated B cell activates lyn kinase, CD19, Akt, IkB, and NFkB. The signaling pathway activated by β2AR and CD86 engagement cause an increase in the coactivator protein OCA-B and the transcription factor Oct-2, respectively. OCA-B and Oct-2 interact and translocate to the nucleus and bind to the 3'IgH-enhancer to increase the rate of IgG1 transcription, which increases the level of IgG1 produced by the cell. *From Sanders, V.M., 2012. The beta2-adrenergic receptor on T and B lymphocytes: do we understand it yet? Brain Behav. Immun. 26, 195–200.*

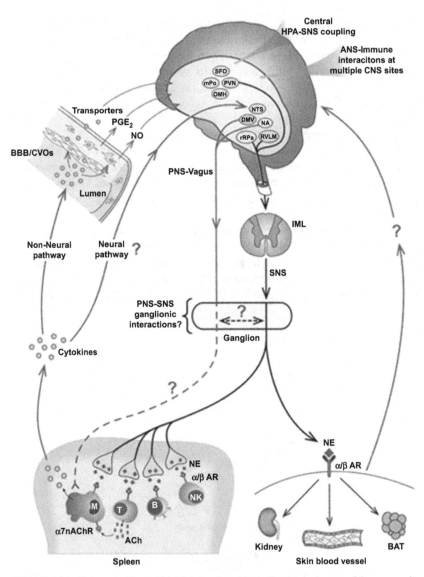

FIGURE 3.31 Schematic diagram highlighting the critical roles that both arms of the autonomic nervous system (i.e., sympathetic nervous system and parasympathetic nervous system) play in regulating central neural and peripheral neuroimmune interactions. Abbreviations: α7nACh, α7 nicotinic acetylcholine receptor; ANS, autonomic nervous system; ARs, adrenergic receptors; B, B cell; BAT, brown adipose tissue; BBB, blood–brain barrier; CNS, central nervous system; CVOs, circumventricular organs; DMH, dorsomedial hypothalamus; DMV, dorsal motor nucleus of the vagus; HPA, hypothalamic–pituitary–adrenal axis; IML, intermediolateral nucleus; M, macrophage; MPO, medial preoptic nucleus; NA, nucleus ambiguous; NE, norepinephrine; NK, natural killer cell; NO, nitric oxide; Ach, acetylcholine; NTS, nucleus tractus solitarius; PGE2, prostaglandin E2; PNS, parasympathetic nervous system; PVN, paraventicular nucleus; rRPa, rostral raphe pallidus; RVLM, rostral ventral lateral medulla; SFO, subfornical organ; SNS, sympathetic nervous system; T, T cell (memory type). *From Kenney, M.J., Ganta, C.K., 2014. Autonomic nervous system and immune system interactions. Comp. Physiol. 4, 1177–1200.*

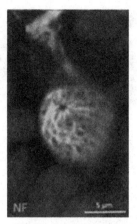

FIGURE 3.32   TNC (thymic nurse cell) closely surrounded by a neural meshwork in the thymic cortex. *From Wülfing, C., Schuran, F.A., Urban, J., Oehlmann, J., Günther, H.S., 2018. Neural architecture in lymphoid organs: hard-wired antigen presenting cells and neurite networks in antigen entrance areas. Immun. Inflamm. Dis. 6, 354–370.*

the number of naïve B cell switching to the production of IgG1, but to the clone of B cells stimulated originally.

Adequate evidence leads us to the firm conclusion that the information for regulating the processes of innate and acquired immunity in vertebrates, both in health and disease, flows from the central nervous system to the immune system. Both the primary (thymus and bone marrow) and secondary (spleen and lymph nodes) lymphoid organs receive sympathetic nervous fibers. The innervation takes place not only at the organ and tissue levels but also deeper at the level of individual immune cells, lymphocytes, macrophages, and dendritic cells (Fig. 3.32).

The acquired immune system and the cellular innate immunity have in common the use of specific receptors to bind the pathogen, but in clear distinction from each other, the cell receptors in the case of innate immune system are inherited, while the receptors of the adaptive immune system are not inherited but are acquired during the life time (Chavan et al., 2017) via an epigenetic mechanism.

A general characteristic of the evolution of immune response in Animalia is a negative correlation with the regenerative capability in animal taxa; the lower they stand on the evolutionary ladder, the higher is their regeneration capability and the simpler is their immune system and the reverse (Peiris et al., 2014) (Fig. 3.33), but the phenomenon remains basically unexplained.

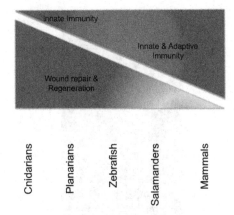

FIGURE 3.33 Inverse correlation between immune system complexity and regenerative capacity. With increasing complexity of the immune system, the regenerative capacity of the organism is decreased. In some invertebrate species without an adaptive immune system and in salamanders with a more complex immune system, scarless repair occurs. In contrast, mammals tend to have scar-forming injury repair and reduced regenerative capacity (Fig. 3.33). *From Peiris, T.H., Hoyer, K.K., Oviedo, N.J., 2014. Innate immune system and tissue regeneration in planarians: an area ripe for exploration. Semin. Immunol. 26, 295−302.*

References

Albalat, R., Martí-Solans, J., Cañestro, C., 2012. DNA methylation in amphioxus: from ancestral functions to new roles in vertebrates. Brief. Funct. Genomics 11, 142−155.

Allen, S.E., Darnell, R.B., Lipscombe, D., 2010. The neuronal splicing factor Nova controls alternative splicing in N-type and P-type Ca$_V$2 calcium channels. Channels 4 (6), 483−489, 2010 Nov-Dec.

Araújo, M.A., Marques, T.E., Octacílio-Silva, S., Arroxelas-Silva, C.L., Pereira, M.G., Peixoto-Santos, J.E., et al., 2016. Identification of microRNAs with dysregulated expression in status epilepticus induced epileptogenesis. PLoS One 11 (10), e0163855, 2016 Oct 3.

Arresta, E., Bernardini, S., Gargioli, C., Filoni, S., Cannata, S.M., 2005. Lens-forming competence in the epidermis of *Xenopus laevis* during development. J. Exp. Zool. A Comp. Exp. Biol. 303, 1−12.

Arrigo, A.P., Simon, S., 2010. Expression and functions of heat shock proteins in the normal and pathological mammalian eye. Curr. Mol. Med. 10, 776−793.

Artieri, C.G., Haerty, W., Singh, R.S., 2009. Ontogeny and phylogeny: molecular signatures of selection, constraint, and temporal pleiotropy in the development of *Drosophila*. BMC Biol. 7, 42.

Artyukhin, A.B., Yim, J.J., Srinivasan, J., Izrayelit, Y., Bose, N., von Reuss, S.H., et al., 2013. Succinylated octopamine ascarosides and a new pathway of biogenic amine metabolism in *Caenorhabditis elegans*. J. Biol. Chem. 288, 18778−18783.

Ast, G., 2004. How did alternative splicing evolve? Nat. Rev. Genet. 5, 773−782.

Bakirtzi, K., Hatziapostolou, M., Karagiannides, I., Polytarchou, C., Jaeger, S., Iliopoulos, D., Pothoulakis, C., 2011. Neurotensin signaling activates microRNAs-21 and −155 and Akt, promotes tumor growth in mice, and is increased in human colon tumors. Gastroenterology 141, 1749–1761.

Bakirtzi, K., Law, I.K.M., Xue, X., Iliopoulos, D., Shah, Y.M., Pothoulakis, C., 2016. Neurotensin promotes the development of colitis and intestinal angiogenesis via hif-1α–mir-210 signaling. J. Immunol. 196, 4311–4321.

Bashirullah, A., Pasquinelli, A.E., Kiger, A.A., Perrimon, N., Ruvkun, G., Thummel, C.S., 2003. Coordinate regulation of small temporal RNAs at the onset of *Drosophila* metamorphosis. Dev. Biol. 259, 1–8.

Baxter, R.H., Chang, C.I., Chelliah, Y., Blandin, S., Levashina, E.A., Deisenhofer, J., 2007. Structural basis for conserved complement factor-like function in the antimalarial protein TEP1. Proc. Natl. Acad. Sci. U.S.A. 104, 11615–11620.

Baxter, R.H.G., Contet, A., Krueger, K., 2017. Arthropod innate immune systems and vector-borne diseases. Biochemistry 56, 907–918.

Beech, R.D., Leffert, J.J., Lin, A., Hong, K.A., Hansen, J., Umlauf, S., et al., 2014. Stress-related alcohol consumption in heavy drinkers correlates with expression of miR-10a, miR-21, and components of the TAR-RNA-binding protein-associated complex. Alcohol Clin. Exp. Res. 38, 2743–2753.

Behesti, H., Papaioannou, V.E., Sowden, J.C., 2009. Loss of Tbx2 delays optic vesicle invagination leading to small optic cups. Dev. Biol. 333, 360–372.

Berke, J.D., Sgambato, V., Zhu, P.P., Lavoie, B., Vincent, M., Krause, M., Hyman, S.E., 2001. Dopamine and glutamate induce distinct striatal splice forms of Ania-6, an RNA polymerase II-associated cyclin. Neuron 32, 277–287.

Bethke, A., Fielenbach, N., Wang, Z., Mangelsdorf, D.J., Antebi, A., 2009. Nuclear hormone receptor regulation of micrornas controls developmental progression. Science 324, 95–98.

Bhat-Nakshatri, P., Wang, G., Collins, N.R., Thomson, M.J., Geistlinger, T.R., Carroll, J.S., et al., 2009. Estradiol-regulated microRNAs control estradiol response in breast cancer cells. Nucleic Acids Res. 37, 4850–4861.

Blaze, J., Roth, T.L., 2017. Caregiver maltreatment causes altered neuronal DNA methylation in female rodents. Dev. Psychopathol. 29, 477–489.

Borck, P.C., Batista, T.M., Vettorazzi, J.F., Soares, G.M., Lubaczeuski, C., Guan, D., 2018. Nighttime light exposure enhances Rev-erbα-targeting microRNAs and contributes to hepatic steatosis. Metabolism 85, 250–258.

Borrelli, E., Nestler, E.J., Allis, C.D., Sassone-Corsi, P., 2008. Decoding the epigenetic language of neuronal plasticity. Neuron 60, 961–974.

Bråte, J., Neumann, R.S., Fromm, B., Haraldsen, A.A.B., Tarver, J.E., Suga, H., et al., 2018. Unicellular origin of the animal MicroRNA machinery. Curr. Biol. 28, 3288–3295.e5.

Braunschweig, U., Barbosa-Morais, N.L., Pan, Q., Nachman, E.N., Alipanahi, B., Gonatopoulos-Pournatzis, T., Frey, B., Irimia, M., Blencowe, B.J., 2014. Widespread intron retention in mammals functionally tunes transcriptomes. Genome Res. 24, 1774–1786.

Bredy, T.W., Wu, H., Crego, C., Zellhoefer, J., Sun, Y.E., Barad, M., 2007. Histone modifications around individual BDNF gene promoters in prefrontal cortex are associated with extinction of conditioned fear. Learn. Mem. 14, 268–276.

Brooks, A.N., Yang, L., Duff, M.O., Hansen, K.D., Park, J.W., Dudoit, S., et al., 2011. Conservation of an RNA regulatory map between *Drosophila* and mammals. Genome Res. 21, 193–202.

Buckanovich, R.J., Yang, Y.Y., Darnell, R.B., 1996. The onconeural antigen Nova-1 is a neuron-specific RNA-binding protein, the activity of which is inhibited by paraneoplastic antibodies. J. Neurosci. 16, 1114–1122.

Cabej, N.R., 2014. On the origin of information in epigenetic structures in metazoans. Med. Hypotheses 83, 378–386.

Cabej, N.R., 2018. Epigenetic Principles of Evolution. Academic Press, London-San Diego-Cambridge, MA- Oxford, p. 355.

Cannata, S.M., Bernardini, S., Filoni, S., Gargioli, C., 2008. The optic vesicle promotes cornea to lens transdifferentiation in larval *Xenopus laevis*. J. Anat. 212, 621–626.

Cardoso, K.C., Da Silva, M.J., Costa, G.G., Torres, T.T., Del Bem, L.E., et al., 2010. A transcriptomic analysis of gene expression in the venom gland of the snake *Bothrops alternatus* (urutu). BMC Genomics 11, 605.

Carletti, M.Z., Fiedler, S.D., Christenson, L.K., 2010. MicroRNA 21 blocks apoptosis in mouse periovulatory granulosa cells. Biol. Reprod. 83, 286–295.

Carmona, L.M., Schatz, D.G., 2017. New insights into the evolutionary origins of the recombination-activating gene proteins and V(D)J recombination. FEBS J. 284, 1590–1605.

Catterall, W.A., 2000. Structure and regulation of voltage-gated Ca^{2+} channels. Annu. Rev. Cell Dev. Biol. 16, 521–555.

Christensen, T.A., Itagaki, H., Teal, P.E., Jasensky, R.D., Tumlinson, J.H., Hildebrand, J.G., 1991. Innervation and neural regulation of the sex pheromone gland in female *Heliothis* moths. Proc. Natl. Acad. Sci. U.S.A. 88, 4971–4975.

Champagne, F.A., 2008. Epigenetic mechanisms and the transgenerational effects of maternal care. Front. Neuroendocrinol. 29, 386–397.

Chavan, S.S., Pavlov, V.A., Tracey, K.J., 2017. Mechanisms and therapeutic relevance of neuro-immune communication. Immunity 46, P927–P942.

Chen, K., Rajewsky, N., 2007. The evolution of gene regulation by transcription factors and microRNAs. Nature Reviews 8, 93–103.

Chen, M., Manley, 2009. Mechanisms of alternative splicing regulation: insights from molecular and genomics approaches. Nat. Rev. Mol. Cell Biol. 10, 741–754.

Cheng, H.Y., Papp, J.W., Varlamova, O., Dziema, H., Russell, B., Curfman, J.P., et al., 2007. microRNA modulation of circadian-clock period and entrainment. Neuron 54, 813–829.

Chiaramonte, M., Russo, R., 2015. The echinoderm innate humoral immune response. Ital. J. Zool. 82, 300–308.

Christodoulou, F., Raible, F., Tomer, R., Simakov, O., Trachana, K., Klaus, S., et al., 2010. Ancient animal microRNAs and the evolution of tissue identity. Nature 463, 1084–1088.

Chute, C.D., Coyle, V., Zhang, Y.K., Rayes, D., Choi, H.J., Srinivasan, J., et al., 2018. Co-option of Neurotransmitter Signaling for Inter-organismal Communication in *C. elegans* bioRxiv 275693.

Cramer, J.M., Pohlmann, D., Gomez, F., Mark, L., Kornegay, B., Hall, C., 2017. Methylation specific targeting of a chromatin remodeling complex from sponges to humans. Sci. Rep. 7. Article number: 40674.

Cui, D., Xu, X., 2018. DNA methyltransferases, DNA methylation, and age-associated cognitive function. Int. J. Mol. Sci. 19, 1315.

Cui, Q., Yu, Z., Purisima, E.O., Wang, E., 2007. MicroRNA regulation and interspecific variation of gene expression. Trends Genet. 23, 372–375.

Cvekl, A., Zhao, Y., McGreal, R., Xie, Q., Gu, X., Zheng, D., 2017. Evolutionary origins of Pax6 control of crystallin genes. Genome Biol. Evol. 9, 2075–2092.

Day, J.J., Childs, D., Guzman-Karlsson, M.C., Kibe, M., Moulden, J., Song, E., Tahir, A., Sweatt, J.D., 2013. DNA methylation regulates associative reward learning. Nat. Neurosci. 16, 1445–1452.

Delsuc, F., Brinkmann, H., Chourrout, D., Philippe, H., 2006. Tunicates and not cephalochordates are the closest living relatives of vertebrates. Nature 439, 965–968.

De Mulder, K., Berezikov, E., 2010. Tracing the evolution of tissue identity with microRNAs. Genome Biol. 11, 111.

de Rie, D., Abugessaisa, I., Alam, T., Arner, E., Arner, P., Ashoor, H., et al., 2017. An integrated expression atlas of miRNAs and their promoters in human and mouse. Nat. Biotechnol. 35, 872–878.

Denlinger, D.L., 2002. Regulation of diapause. Annu. Rev. Entomol. 47, 93–122.

Desvignes, T., Batzel, P., Sydes, J., Eames, B.F., Postlethwait, J.H., 2019. miRNA analysis with Prost! reveals evolutionary conservation of organ-enriched expression and post-transcriptional modifications in three-spined stickleback and zebrafish. Sci. Rep. 9, 3913.

Ding, X., Liu, S., Tian, M., Zhang, W., Zhu, T., Li, D., et al., 2017. Activity-induced histone modifications govern Neurexin-1 mRNA splicing and memory preservation. Nat. Neurosci. 20, 690–699.

Dinh, C., Farinholt, T., Hirose, S., Zhuchenko, O., Kuspa, A., 2018. Lectins modulate the microbiota of social amoebae. Science 361, 402–406.

Doty, K.A., Wilburn, D.B., Bowen, K.E., Feldhoff, P.W., Feldhoff, R.C., 2016. Co-option and evolution of non-olfactory proteinaceous pheromones in a terrestrial lungless salamander. J. Proteomics 135, 101–111.

Eacker, S.M., Keuss, M.J., Berezikov, E., Dawson, V.L., Dawson, T.M., 2011. Neuronal activity regulates hippocampal miRNA expression. PLoS One 6 (10), e25068.

Ebert, M.S., Sharp, P.A., 2012. Roles for MicroRNAs in conferring robustness to biological processes. Cell 149, 505–524.

Egger, B., Steinke, D., Tarui, H., De Mulder, K., Arendt, D., Borgonie, G., et al., 2009. To be or not to be a flatworm: the acoel controversy. PLoS One 4, e5502.

Erclik, T., Hartenstein, V., Lipshitz, H.D., McInnes, R.R., 2008. Conserved role of the *Vsx* genes supports a monophyletic origin for bilaterian visual systems. Curr. Biol. 18, 1278–1287.

Erclik, T., Hartenstein, V., McInnes, R.R., Lipshitz, H.D., 2009. Eye evolution at high resolution: the neuron as a unit of homology. Dev. Biol. 332, 70–79.

Flajnik, M.F., Kasahara, M., 2010. Origin and evolution of the adaptive immune system: genetic events and selective pressures. Nat. Rev. Genet. 11, 47–59.

Feng, S., Cokus, S.J., Zhang, X., Chen, P.-Y., Bostick, M., Goll, M.G., et al., 2010. Conservation and divergence of methylation patterning in plants and animals. Proc. Natl. Acad. Sci. U.S.A. 107, 8689–8694.

Fiedler, S., Christenson, L., 2007. LH/hCG induced expression of microRNAs in murine granulosa cells during the periovulatory period. Biol. Reprod. 77. 192-b-193.

Fiedler, S.D., Carletti, M.Z., Hong, X., Christenson, L.K., 2008. Hormonal regulation of MicroRNA expression in periovulatory mouse mural granulosa cells. Biol. Reprod. 79, 1030–1037.

Fiore, R., Khudayberdiev, S., Christensen, M., Siegel, G., Flavell, S.W., et al., 2009. Mef2-mediated transcription of the miR379-410 cluster regulates activity-dependent dendritogenesis by fine-tuning Pumilio2 protein levels. EMBO J. 28, 697–710.

Franchi, N., Ballarin, L., 2017. Immunity in protochordates: the tunicate perspective. Front. Immunol. 8, 674.

Friedman, R.C., Farh, K.K., Burge, C.B., Bartel, D.P., 2009. Most mammalian mRNAs are conserved targets of microRNAs. Genome Res. 19, 92–105.

Funk, D.H., Sweeney, B.W., Jackson, J.K., 2010. Why stream mayflies can reproduce without males but remain bisexual: a case of lost genetic variation. J. North Am. Benthol. Soc. 29, 1258−1266.

Furuta, Y., Hogan, B.L.M., 1998. BMP4 is essential for lens induction in the mouse embryo. Genes Dev. 12, 3764−3775.

Gaiti, F., Jindrich, K., Fernandez-Valverde, S.L., Roper, K.E., Degnan, B.M., Tanurdžić, M., 2017. Landscape of histone modifications in a sponge reveals the origin of animal cis-regulatory complexity. eLife 6, e22194.

Galoian, K., Galoian, K., 2017. Epigenetic control of cancer by neuropeptides (Review). Biomedical Reports 6, 3−7.

Gao, L., Li, A., Li, N., Liu, X., Deng, H., Zhao, B., Pang, Q., 2017. Innate and intrinsic immunity in planarians. ISJ 14, 443−452.

Gebremedhn, S., Salilew-Wondim, D., Ahmad, I., Sahadevan, S., Hossain, M.M., Hoelker, M., et al., 2015. MicroRNA expression profile in bovine granulosa cells of preovulatory dominant and subordinate follicles during the late follicular phase of the estrous cycle. PLoS One 10 (5), e0125912.

Gehring, W.J., 1996. The master control gene for morphogenesis and evolution of the eye. Genes Cells 1, 11−15.

Gilbert, J.J., 2016. Non-genetic polymorphisms in rotifers: environmental and endogenous controls, development, and features for predictable or unpredictable environments. Biol. Rev. Camb. Philos. Soc. 792, 964−992.

Goll, M.G., Bestor, T.H., 2005. Eukaryotic cytosine methyltransferases. Annu. Rev. Biochem. 74, 481−514.

Goss, R.J., 1969. Principles of Regeneration. Academic Press, New York; London, p. 80.

Granjon, A., Gustin, M.P., Rieusset, J., Lefai, E., Meugnier, E., Güller, I., et al., 2009. The microRNA signature in response to insulin reveals its implication in the transcriptional action of insulin in human skeletal muscle and the role of a sterol regulatory element−binding protein-1c/myocyte enhancer factor 2C pathway. Diabetes 58, 2555−2564.

Graw, J., 1996. Genetic aspects of embryonic eye development in vertebrates. Dev. Genet. 18, 181−197.

Griggs, E.M., Young, E.J., Rumbaugh, G., Miller, C.A., 2013. microRNA-182 regulates amygdala-dependent memory formation. J. Neurosci. 33, 1734−1740.

Grimson, A., Srivastava, M., Fahey, B., Woodcroft, B.J., Chiang, H.R., King, N., 2008. Early origins and evolution of microRNAs and Piwi-interacting RNAs in animals. Nature 455, 1193−1197.

Grossman, H., Chuderland, D., Ninio-Many, L., Hasky, N., Kaplan-Kraicer, R., Shalg, R., 2015. A novel regulatory pathway in granulosa cells, the LH/human chorionic gonadotropin-microRNA-125a-3p-Fyn pathway, is required for ovulation. FASEB J. 29, 3206−3216.

Grün, D., Wang, Y.-L., Langenberger, D., Gunsalus, K.C., Rajewsky, N., 2005. microRNA target predictions across seven Drosophila species and comparison to mammalian targets. PLoS Comput. Biol. 1 (1), e13.

Guan, Z., Giustetto, M., Lomvardas, S., Kim, J.H., Miniaci, M.C., Schwartz, J.H., et al., 2002. Integration of long-term-memory-related synaptic plasticity involves bidirectional regulation of gene expression and chromatin structure. Cell 111, 483−493.

Guo, J.U., Su, Y., Zhong, C., Ming, G-l., Song, H., 2011. Emerging roles of TET proteins and 5-hydroxymethylcytosines in active DNA demethylation and beyond. Cell Cycle 10 (16), 2662−2668.

Guo, L., Xu, J., Qi, J., Zhang, L., Wang, J., Liang, J., Qian, N., Zhou, H., Wei, L., Deng, L., 2013. MicroRNA-17−92a upregulation by estrogen leads to Bim targeting and inhibition of osteoblast apoptosis. J. Cell Sci. 126, 978−988.

Halder, G., Callaerts, P., Gehring, W.J., 1995. Induction of ectopic eyes by targeted expression of the Eyeless gene in *Drosophila*. Science 267, 1788−1792.

Hamon, M.A., Cossart, P., 2008. Histone modifications and chromatin remodeling during bacterial infections. Cell Host Microbe 4, P100−P109.

Hansen, K.L.F., 2015. MiR-132 as a Dynamic Regulator of Neuronal Structure and Cognitive Capacity. Dissert. Ohio State University.

Harlin-Cognato, A., Hoffman, E.A., Jones, A.G., 2006. Gene cooption without duplication during the evolution of a male-pregnancy gene in pipefish. Proc. Natl. Acad. Sci. U.S.A. 103, 19407−19412.

Hattori, D., Millard, S.S., Wojtowicz, W.M., Zipursky, S.L., 2008. Dscam-mediated cell recognition regulates neural circuit formation. Annu. Rev. Cell Dev. Biol. 24, 597−620.

Heimberg, A.M., Sempere, L.F., Moy, V.N., Donoghue, P.C.J., Peterson, K.J., 2008. MicroRNAs and the advent of vertebrate morphological complexity. Proc. Natl. Acad. Sci. U.S.A. 105, 2946−2950.

Hermey, G., Blüthgen, N., Kuhl, D., 2017. Neuronal activity-regulated alternative mRNA splicing. Int. J. Biochem. Cell Biol. 91 (Part B), 184−193.

Hertel, J., Lindemeyer, M., Missal, K., Fried, C., Tanzer, A., Flamm, C., et al., 2006. The expansion of the metazoan microRNA repertoire. BMC Genomics 7, 25.

Heyward, J.D., Sweatt, J.D., 2015. DNA methylation in memory formation: emerging insights. Neuroscientist 21, 475−489.

Hilliard, M.A., Apicella, A.J., Kerr, R., Suzuki, H., Bazzicalupo, P., Schafer, W.R., 2005. In vivo imaging of *C. elegans* ASH neurons: cellular response and adaptation to chemical repellents. EMBO J. 24, 63−72.

Holliday, R., 2002. Epigenetics comes of age in the twenty-first century. J. Genet. 81, 1−4.

Hu, Z., Shen, W.J., Kraemer, F.B., Azhar, S., 2012a. MicroRNAs 125a and 455 repress lipoprotein-supported steroidogenesis by targeting scavenger receptor class B type I in steroidogenic cells. Mol. Cell. Biol. 32, 5035−5045.

Hu, Z., Shen, W.J., Kraemer, F.B., Azhar, S., 2012b. MicroRNAs 125a and 455 repress lipoprotein-supported steroidogenesis by targeting scavenger receptor class B type I in steroidogenic cells. Mol. Cell. Biol. 32, 5035−5045.

Hu, Z., Scott, H.S., Qin, G., Zheng, G., Chu, X., Xie, L., et al., 2015. Revealing missing human protein isoforms based on Ab initio prediction, RNA-seq and proteomics. Sci. Rep. 5. Article number: 10940.

Hu, Z., Shen, W.-J., Cortez, Y., Tang, X., Liu, L.-F., Kraemer, F.B., Azhar, S., 2013. Adrenals from ACTH, 17α-E2 and dexamethasone treated rats exhibited miRNA profiles distinct from control animals. PLoS One 8 (10), e78040.

Igaz, I., 2015. Changes in Expression of Circulating microRNAs after Hormone Administration and Their Potential Biological Relevance. Ph.D. Theses. Semmelweis University, Budapest.

Iida, H., Ishii, Y., Kondoh, H., 2017. Intrinsic lens potential of neural retina inhibited by Notch signaling as the cause of lens transdifferentiation. Dev. Biol. 421, 118−125.

Iijima, T., Wu, K., Witte, H., Hanno-Iijima, Y., Glatter, T., Richard, S., Scheiffele, P., 2011. SAM68 regulates neuronal activity-dependent alternative splicing of neurexin-1. Cell 147, 1601−1614.

Iijima, T., Hidaka, C., Iijima, Y., 2016. Spatio-temporal regulations and functions of neuronal alternative RNA splicing in developing and adult brains. Neurosci. Res. 109, 1−8.

Irimia, M., Denuc, A., Burguera, D., Somorjai, I., Martín-Durán, J.M., Genikhovich, G., et al., 2011. Stepwise assembly of the Nova-regulated alternative splicing network in the vertebrate brain. Proc. Natl. Acad. Sci. U.S.A. 108, 5319—5324.

Iwaki, D., Kanno, K., Takahashi, M., EndoY, Matsushita, M., Fujita, T., 2011. The role of Mannose-binding-lectin-associated Serine protease-3 in activation of the alternative complement pathway. J. Immunol. 187, 3751—3758.

Janeway Jr., C.A., Travers, P., Walport, M., Shlomchik, M., 2001a. The complement system and innate immunity. In: Immunobiology: The ImmuneSystem in Health and Disease, fifth ed. Garland Science, New York. Available from: https://www.ncbi.nlm.nih.gov/books/NBK27100.

Janeway Jr., C.A., Travers, P., Walport, M., Shlomchik, M., 2001b. Immunobiology: The Immune System in Health and Disease, fifth ed. Garland Science, New York, p. 601.

Janssenswillen, S., Vandebergh, W., Treer, D., Willaert, B., Maex, M., Van Bocxlaer, I., Bossuyt, F., 2015. Origin and diversification of a salamander sex pheromone system. Mol. Biol. Evol. 32, 472—480.

Jarome, T.J., Lubin, F.D., November 2014. Epigenetic mechanisms of memory formation and reconsolidation. Neurobiol. Learn. Mem. 0, 116—127.

Jelen, N., Ule, J., Živin, M., Darnell, R.B., 2007. Evolution of nova-dependent splicing regulation in the brain. PLoS Genet. 3 (10), e173.

Jeltsch, A., 2010. Phylogeny of methylomes. Science 328, 837—838.

Jeng, S.F., Rau, C.S., Liliang, P.C., Wu, C.J., Lu, T.H., Chen, Y.C., Lin, C.J., Hsieh, C.H., 2009. Profiling muscle-specific microRNA expression after peripheral denervation and re-innervation in a rat model. J. Neurotrauma 26, 2345—2353.

Jeppsson, S., Srinivasan, S., Chandrasekharan, B., 2017. Neuropeptide Y (NPY) promotes inflammation-induced tumorigenesis by enhancing epithelial cell proliferation. Am. J. Physiol. Gastrointest. Liver Physiol. 312, G103—G111.

Jia, K., Albert, P.S., Riddle, D.L., 2002. DAF-9, a cytochrome P450 regulating *C. elegans* larval development and adult longevity. Development 129, 221—231.

Junqueira-de-Azevedo, I.L.M., Bastos, C.M.V., Ho, P.L., Luna, M.S., Yamanouye, N., Casewell, N.R., 2015. Venom-related transcripts from *Bothrops jararaca* tissues provide novel molecular insights into the production and evolution of snake venom. Mol. Biol. Evol. 32, 754—766.

Kamm, K., Schierwater, B., DeSalle, R., 2019. Innate immunity in the simplest animals—placozoans. BMC Genomics 20, 5.

Kapitonov, V.V., Koonin, E.V., 2015. Evolution of the RAG1-RAG2 locus: both proteins came from the same transposon. Biol. Direct 10 (2015), 20.

Katayama, Y., Battista, M., Kao, W.-M., Hidalgo, A., Peired, A.J., Thomas, S.A., Frenette, P.S., 2006. Signals from the sympathetic nervous system regulate hematopoietic stem cell egress from bone marrow. Cell 124, 407—421.

Katsuura, S., Kuwano, Y., Yamagishi, N., Kurokawa, K., Kajita, K., Akaike, Y., et al., 2012. MicroRNAs miR-144/144* and miR-16 in peripheral blood are potential biomarkers for naturalistic stress in healthy Japanese medical students. Neurosci. Lett. 516, 79—84.

Keller, T.E., Han, P., Yi, S.V., 2016. Evolutionary transition of promoter and gene body DNA methylation across invertebrate—vertebrate boundary. Mol. Biol. Evol. 33, 1019—1028.

Kelly, T.K., Ahmadiantehrani, S., Blattler, A., London, S.E., 2018. Epigenetic regulation of transcriptional plasticity associated with developmental song learning. Proc. R. Soc. B 285, 20180160.

Kenney, M.J., Ganta, C.K., 2014. Autonomic nervous system and immune system interactions. Comp. Physiol. 4, 1177—1200.

Kim, E., Magen, A., Ast, G., 2007. Different levels of alternative splicing among eukaryotes. Nucleic Acids Res. 35, 125–131.

Klimova, L., Kozmik, Z., 2014. Stage-dependent requirement of neuroretinal Pax6 for lens and retina development. Development 141, 1292–1302.

Koberstein, J.N., Poplawski, S.G., Wimmer, M.E., Porcari, G., Kao, C., Gomes, B., et al., 2018. Learning-dependent chromatin remodeling highlights noncoding regulatory regions linked to autism, 2018 Jan 16 Sci. Signal. 11 (513). eaan6500.

Korotkov, A., Mills, J.D., Gorter, J.A., van Vliet, E.A., Aronica, E., 2017. Systematic review and meta-analysis of differentially expressed miRNAs in experimental and human temporal lobe epilepsy. Sci. Rep. 7, 11592.

Kosik, K.S., 2009. Exploring the early origins of the synapse by comparative genomics. Biol. Lett. 5, 108–111.

Krishnan, J., Rohner, N., 2017. Cavefish and the basis for eye loss. Philos. Trans. R. Soc. Lond. B Biol. Sci. 372, 20150487.

Krol, J., Busskamp, V., Markiewicz, I., Stadler, M.B., Ribi, S., Richter, J., et al., 2010. Characterizing light-regulated retinal MicroRNAs reveals rapid turnover as a common property of neuronal MicroRNAs. Cell 141, 618–631.

Kye, M.J., Neveu, P., Lee, Y.-S., Zhou, M., Steen, J.A., Sahin, M., et al., 2011. NMDA mediated contextual conditioning changes miRNA expression. PLoS One 6 (9), e24682.

Laird, D.J., De Tomaso, A.W., Cooper, M.D., Weissman, I.L., 2000. 50 million years of chordate evolution: seeking the origins of adaptive immunity. Proc.Natl. Acad. Sci. 97, 6924–6926.

Laulan, A., Lestage, J., Chateaureynaud-Duprat, P., Fontaine, M., 1983. Some substances from the coelomic fluid of *Lumbricus terrestris* possess functions common to certain components of the human complement. Ann. Immunol. 134, 223–232.

Lee, R.C., Feinbaum, R.L., Ambros, V., 1993. The *C. elegans* heterochronic gene *lin-4* encodes small RNAs with antisense complementarity to *lin-14*. Cell 75, 843–854.

Lee, Y., Rio, D.C., 2015. Mechanisms and regulation of alternative pre-mRNA splicing. Annu. Rev. Biochem. 84, 291–323.

Lenski, R.E., Barrick, J.E., Ofria, C., 2006. Balancing robustness and evolvability. PLoS Biol. 4, 2190–2192.

Leung, A.K., Sharp, P.A., 2007. microRNAs: a safeguard against turmoil? Cell 130, 581–585.

Leung, A.K., Sharp, P.A., 2010. MicroRNA functions in stress responses. Mol. Cell 40, P205–P215.

Levenson, J.M., O'Riordan, K.J., Brown, K.D., Trinh, M.A., Molfese, D.L., Sweatt, J.D., 2004. Regulation of histone acetylation during memory formation in the hippocampus. J. Biol. Chem. 279, 40545–40559.

Li, Q., Lee, J.-A., Black, D.L., 2007. Neuronal regulation of alternative pre-mRNA splicing. Nat. Rev. Neurosci. 8, 819–831.

Lin, L., Shen, S., Jiang, P., Sato, S., Davidson, B.L., Xing, Y., 2010. Evolution of alternative splicing in primate brain transcriptomes. Hum. Mol. Genet. 19, 2958–2973.

Lister, R., Pelizzola, M., Dowen, R.H., Hawkins, R.D., Hon, G., Tonti-Filippini, J., 2009. Human DNA methylomes at base resolution show widespread epigenomic differences. Nature 462, 315–322.

Liu, L., Laufer, H., 1996. Isolation and characterization of sinus gland neuropeptides characterization of sinus gland neuropeptides with bothmandibular organ inhibiting and hyperglycemiceffects from the spider crab, *Libinia emarginata*. Arch. Insect Biochem. Physiol. 32, 375–385.

Loker, E.S., 2012. Macroevolutionary immunology: a role for immunity in the diversification of animal life. Front. Immunol. 3, 25.

Londin, E., Loher, P., Telonis, A.G., Quann, K., Clark, P., Jing, Y., et al., 2015. Analysis of 13 cell types reveals evidence for the expression of numerous novel primate- and tissue-specific microRNAs. Proc. Natl. Acad. Sci. U.S.A. 112, E1106–E1115.

LaVoie, H.A., 2005. Epigenetic control of ovarian function: the emergingrole of histone modifications. Mol. Cell. Endocrinol. 243, 12–18.

Luna, M.S.A., Hortencio, T.M.A., Ferreira, Z.S., Yamanouye, N., 2009. Sympathetic outflow activates the venom gland of the snake *Bothrops jararaca* by regulating the activation of transcription factors and the synthesis of venom gland proteins. J. Exp. Biol. 212, 1535–1543.

Luna, M.S., Valente, R.H., Perales, J., Vieira, M.L., Yamanouye, N., 2013. Activation of *Bothrops jararaca* snake venom gland and venom production: a proteomic approach. J. Proteomics 94, 460–472.

Lynch, M., Conery, J.S., 2000. The evolutionary fate and consequences of duplicate genes. Science 290, 1151–1155.

Ma, D.K., Jang, M.-H., Guo, J.U., Kitabatake, Y., Chang, M-l., Pow-anpongkul, N., et al., 2009. Neuronal activity–induced Gadd45b promotes epigenetic DNA demethylation and adult neurogenesis. Science 323, 1074–1077.

Magnusson, C., Libelius, R., Tågerud, S., 2003. Nogo (Reticulon 4) expression in innervated and denervated mouse skeletal muscle. Mol. Cell. Neurosci. 22, 298–307.

Marden, J.H., 2008. Quantitative and evolutionary biology of alternative splicing: how changing the mix of alternative transcripts affects phenotypic plasticity and reaction norms. Heredity 100, 111–120.

Mathew, R.S., Tatarakis, A., Rudenko, A., Johnson-Venkatesh, E.M., Yang, Y.J., Murphy, E.A., et al., 2016. A microRNA negative feedback loop downregulates vesicle transport and inhibits fear memory. Elife 5, e22467, 2016 Dec 21.

Mauger, O., Lemoine, F., Scheiffele, P., 2016. Targeted intron retention and excision for rapid gene regulation in response to neuronal activity. Neuron 92, 1266–1278.

McLennan, D.A., 2008. The concept of Co-option: why evolution often looks. Miraculous Evo Edu Outreach 1, 247. https://doi.org/10.1007/s12052-008-0053-8.

McManus, C.J., Coolon, J.D., Eipper-Mains, J., Wittkopp, P.J., Graveley, B.R., 2014. Evolution of splicing regulatory networks in *Drosophila*. Genome Res. 24, 786–796.

Meaney, M.J., Szyf, M., 2005. Maternal care as a model for experience-dependent chromatin plasticity? Trends Neurosci. 28, 456–463.

Meister, M., Lemaitre, B., Hoffmann, J.A., 1997. Antimicrobial peptide defense in *Drosophila*. Bioessays 19, 1019–1026.

Meuti, M.E., Bautista-Jimenez, R., Reynolds, J.A., 2018. Evidence that microRNAs are part of the molecular toolkit regulating adult reproductive diapause in the mosquito, *Culex pipiens*. PLoS One 13 (11), e0203015.

Miller, D.J., Hemmrich, G., Ball, E.E., Hayward, D.C., Khalturin, K., Funayama, N., Agata, K., Bosch, T.C., 2007. The innate immune repertoire in cnidaria: ancestral complexity and stochastic gene loss. Genome Biol. 8, R59.

Miller, C.A., Gavin, C.F., White, J.A., Parrish, R.R., Honasoge, A., Yancey, C.R., 2010. Cortical DNA methylation maintains remote memory. Nat. Neurosci. 13, 664–666.

Missler, M., Südhof, T.C., 1998. Neurexins: three genes and 1001 products. Trends Genet. 14, 20–26.

Mollet, I.G., Ben-Dov, C., Felício-Silva, D., Grosso, A.R., Eleutério, P., Alves, R., et al., 2010. Unconstrained mining of transcript data reveals increased alternative splicing complexity in the human transcriptome. Nucleic Acids Res. 38, 4740–4754.

Morlando, M., Ballarino, M., Gromak, N., Pagano, F., Bozzoni, I., Proudfoot, N.J., 2008. Primary microRNA transcripts are processed co-transcriptionally. Nat. Struct. Mol. Biol. 15, 902–909.

Mu, Y., Otsuka, T., Horton, A.C., Scott, D.B., Ehlers, M.D., 2003. Activity-dependent mRNA splicing controls ER export and synaptic delivery of NMDA receptors. Neuron 40, 581–594.

Murphy, C.P., Singewald, N., 2018. Potential of microRNAs as novel targets in the alleviation of pathological fear. Genes Brain Behav. 17 (3), e12427.

Neubauer, E.-F., Poole, A.Z., Neubauer, P., Detournay, O., Tan, K., Davy, S.K., Weis, V.M., 2017. A Diverse Host Thrombospondin-Type-1 Repeat Protein Repertoire Promotes Symbiont Colonization during Establishment of Cnidarian-Dinoflagellate Symbiosis. ELife e24494.

Nilsen, T.W., Graveley, B.R., 2010. Expansion of the eukaryotic proteome by alternative splicing. Nature 463, 457–463.

Nilsson, D.-E., 1996. Eye ancestry: old genes for new eyes. Curr. Biol. 6, 39–42.

Nilsson, D.E., 2009. The evolution of eyes and visually guided behaviour. Philos. Trans. R. Soc. Lond. Ser. B Biol. Sci. 364, 2833–2847.

Nudelman, A.S., DiRocco, D.P., Lambert, T.J., Garelick, M.G., Le, J., Nathanson, N.M., Storm, D.R., 2010. Neuronal activity rapidly induces transcription of the CREB-regulated microRNA-132, *in vivo*. Hippocampus 20, 492–498.

Nunes-Burgos, G.B., Gonçalves, L.R.C., Furtado, M.F.D., Fernandes, W., Nicolau, J., 1993. Alteration of the protein composition of *Bothrops jararaca* venom and venom gland by isoproterenol treatment. Int. J. Biochem. 25, 1491–1496.

O'Brien, J., Hayder, H., Zayed, Y., Peng, C., 2018. Overview of MicroRNA biogenesis, mechanisms of actions, and circulation. Front. Endocrinol. 9 (402).

Osenberg, S., Karten, A., Sun, J., Li, J., Charkowick, S., Felice, C.A., et al., 2018. Activity-dependent aberrations in gene expression and alternative splicing in a mouse model of Rett syndrome. Proc. Natl. Acad. Sci. U.S.A. 115, E5363–E5372.

Palmer, C.A., Watts, R.A., Houck, L.D., Picard, A.L., Arnold, S.J., 2007. Evolutionary replacement of components in a salamander pheromone signaling complex: more evidence for phenotypic molecular decoupling. Evolution 61, 202–215.

Pandya, P.H., Murray, M.E., Pollok, K.E., Renbarger, J.L., 2016. The immune system in cancer pathogenesis: potential therapeutic approaches. J. Immunol. Res. 2016 (4273943).

Park, C.S., Tang, S.J., 2009. Regulation of microRNA expression by induction of bidirectional synaptic plasticity. J. Mol. Neurosci. 38, 50–56.

Pearce, K., Cai, D., Roberts, A.C., Glanzman, D.L., 2017. Role of protein synthesis and DNA methylation in the consolidation and maintenance of long-term memory in Aplysia. Elife 6, e18299.

Peiris, T.H., Hoyer, K.K., Oviedo, N.J., 2014. Innate immune system and tissue regeneration in planarians: an area ripe for exploration. Semin. Immunol. 26, 295–302.

Pestarino, M., Oliveri, D., Parodi, M., Candiani, S., 2007. The amphioxus immune system. ISJ 4, 45–50.

Peterson, K.J., Dietrich, M.R., McPeek, M.A., 2009. MicroRNAs and metazoan macroevolution: insights into canalization, complexity, and the Cambrian explosion. Bioessays 31, 736–747.

Peterson, K.J., Sperling, E.A., 2007. Poriferan ANTP genes: primitively simple or secondarily reduced? Evol. Dev. 9, 405–408.

Piatigorsky, J., Wistow, G., 1991. The recruitment of crystallins: new functions precede gene duplication. Science 252, 1078−1079.

Pinney, S.E., 2014. Mammalian non-CpG methylation: stem cells and beyond. Biology 3, 739−751.

Plachetzki, D.C., Oakley, T.H., 2007. Key transitions during the evolution of animal photo-transduction: novelty, "tree-thinking," co-option, and co-duplication. Integr. Comp. Biol. 47, 759−769.

Polydorides, A.D., Okano, H.J., Yang, Y.Y.L., Stefani, G., Darnell, R.B., 2000. A brain-enriched polypyrimidine tract-binding protein antagonizes the ability of Nova to regulate neuron-specific alternative splicing. Proc. Natl. Acad. Sci. U.S.A. 97, 6350−6355.

Poole, A.Z., Kitchen, S.A., Weis, V.M., April 22, 2016. The role of complement in Cnidarian-dinoflagellate symbiosis and immune challenge in the Sea anemone *Aiptasia pallida*. Front. Microbiol.

Quesnel-Vallières, M., Dargaei, Z., Irimia, M., Woodin, M.A., Blencowe, B.J., Cordes, S.P., 2016. Misregulation of an activity-dependent splicing network as a common mechanism underlying autism spectrum disorders. Mol. Cell 64, 1023−1034.

Reichert, V., Moore, M.J., 2000. Better conditions for mammalian in vitro splicing provided by acetate and glutamate as potassium counterions. Nucleic Acids Res. 28, 416−423.

Rétaux, S., Casane, D., 2013. Evolution of eye development in the darkness of caves: adaptation, drift, or both? EvoDevo 4, 26.

Reynolds, J.A., Nachman, R.J., Denlinger, D.L., 2018. Distinct microRNA and mRNA responses elicited by ecdysone, diapause hormone and a diapause hormone analog at diapause termination in pupae of the corn earworm, *Helicoverpa zea*. Gen. Comp. Endocrinol. Available online 20 September 2018. (in press), Corrected Proof.

Reza, H.M., Yasuda, K., 2004a. Lens differentiation and crystallin regulation: a chick model. Int. J. Dev. Biol. 48, 805−817.

Reza, H.M., Yasuda, K., 2004b. The involvement of neural retina Pax6 in lens fiber differentiation. Dev. Neurosci. 26, 318−327.

Riesgo, A., Farrar, N., Windsor, P.J., Giribet, G., Leys, S.P., 2014. The analysis of eight transcriptomes from all poriferan classes reveals surprising genetic complexity in sponges. Mol. Biol. Evol. 31, 1102−1120.

Riester, A., Issler, O., Spyroglou, A., Rodrig, S.H., Chen, A., Beuschlein, F., 2011. ACTH-dependent regulation of microRNA as endogenous modulators of glucocorticoid receptor expression in the adrenal gland. Endocrinology 153, 212−222.

Rollmann, S.M., Houck, L.D., Feldhoff, R.C., 1999. Proteinaceous pheromone affecting female receptivity in a terrestrial salamander. Science 285, 1907−1909.

Rossi, L., Iacopetti, P., Salvetti, A., 2012. Stem cells and neural signalling: the case of neoblast recruitment and plasticity in low dose X-ray treated planarians. Int. J. Dev. Biol. 56, 135−142.

Rozic, G., Lupowitz, Z., Piontkewitz, Y., Zisapel, N., 2011. Dynamic changes in neurexins' alternative splicing: role of rho-associated protein kinases and relevance to memory formation. PLoS One 6 (4), e18579.

Rusconi, F., Battaglioli, E., 2018. Acute stress-induced epigenetic modulations and their potential protective role toward depression. Front. Mol. Neurosci. 11 (2018), 184.

Salta, E., De Strooper, B., 2018. microRNA-132: a key noncoding RNA operating in the cellular phase of Alzheimer's disease. FASEB J. 31, 424−433.

Salvador, L.M., Park, Y., Cottom, J., Maizels, E.T., Jones, J.C.R., Schillace, R.V., et al., 2001. Follicle-stimulating hormone stimulates protein kinase A-mediated histone H3 phosphorylation and acetylation leading to select gene activation in ovarian granulosa cells. J. Biol. Chem. 276, 40146–40155.

Salzet, M., Tasiemski, A., Cooper, E., 2006. Innate immunity in lophotrochozoans: the annelids. Curr. Pharmaceut. Des. 12, 3043–3050.

Sambandan, S., Akbalik, G., Kochen, L., Rinne, J., Kahlstatt, J., Glock, C., et al., 2017. Activity-dependent spatially localized miRNA maturation in neuronal dendrites. Science 355, 634–637.

Sanders, V.M., 2012. The beta2-adrenergic receptor on T and B lymphocytes: do we understand it yet? Brain Behav. Immun. 26, 195–200.

Santoro, S.W., Dulac, C., 2012. The activity-dependent histone variant H2BE modulates the life span of olfactory neurons. eLife 1 (2012), e00070.

Sarma, S.S.S., Nandini, S., 2007. Small prey size offers immunity to predation: a case study on two species of Asplanchna and three brachionid prey (Rotifera). Hydrobiologia 93, 67–76.

Sgambato, V., Minassian, R., Nairn, A.C., Hyman, S.E., 2003. Regulation of ania-6 splice variants by distinct signaling pathways in striatal neurons. J. Neurochem. 86, 153–164.

Shaham, O., Smith, A.N., Robinson, M.L., Taketo, M.M., Lang, R.A., Ashery-Padan, R., 2009. Pax6 is essential for lens fiber cell differentiation. Development 136, 2567–2578.

Sharma, A., Lou, H., 2011. Depolarization-mediated regulation of alternative splicing. Front. Neurosci. 5, 141.

Shomron, N., Golan, D., Hornstein, E., 2009. An evolutionary perspective of animal MicroRNAs and their targets. J. Biomed. Biotechnol. 2009, 1–9.

Silerová, M., Procházková, P., Josková, R., Josens, G., Beschin, A., De Baetselier, P., Bilej, M., 2006. Comparative study of the CCF-like pattern recognition protein in different Lumbricid species. Dev. Comp. Immunol. 30, 765–771.

Smith, L.K., Shah, R.R., Cidlowski, J.A., 2010. Glucocorticoids modulate microRNA expression and processing during lymphocyte apoptosis. J. Biol. Chem. 285, 36698–36708.

Stark, A., Brennecke, J., Bushati, N., Russell, R.B., Cohen, S.M., 2005. Animal MicroRNAs confer robustness to gene expression and have a significant impact on 3'UTR evolution. Cell 123, 1133–1146.

Stelzer, C.P., 2008. Obligate asex in a rotifer and the role of sexual signals. J. Evol. Biol. 21, 287–293.

Stork, O., Stork, S., Pape, H.-C., Obata, K., 2001. Identification of genes expressed in the amygdala during the formation of fear memory. Learn. Mem. 8, 209–219.

Su, C.-H., Dhananjaya, D., Tarn, W.-Y., 2018. Alternative splicing in neurogenesis and brain development. Front. Mol. Biosci. 5 (2018), 12.

Su, Y., Shin, J., Zhong, C., Wang, S., Roychowdhury, P., Lim, J., et al., 2017. Neuronal activity modifies the chromatin accessibility landscape in the adult brain. Nat. Neurosci. 20, 476–483.

Su, Z., Wang, J., Yu, J., Huang, X., Gu, X., 2006. Evolution of alternative splicing after gene duplication. Genome Res. 16, 182–189.

Teal, P.E.A., Davis, N.T., Meredith, J.A., Christensen, T.A., Hildebrand, J.G., 1999. Role of the ventral nerve cord and terminal abdominal ganglion in the regulation of sex pheromone production in the tobacco budworm (Lepidoptera: noctuidae). Ann. Entomol. Soc. Am. 92, 891–901.

Teal, P.E.A., Tumlinson, J.H., Oberlander, H., 1989. Neural regulation of sex pheromone biosynthesis in *Heliothis* moths. Proc. Natl. Acad. Sci. U.S.A. 86, 2488–2492.

Technau, U., 2008. Small regulatory RNAs pitch in. Nature 455, 1184–1185.

Thein, T., de Melo, J., Zibetti, C., Clark, B.S., Juarez, F., Blackshaw, S., 2016. Control of lens development by Lhx2-regulated neuroretinal FGFs. Development 143, 3994–4002.

Thomson, D.W., Dinger, M.E., 2016. Endogenous microRNA sponges: evidence and controversy. Nat. Rev. Genet. 17, 272–283.

Thurman, R.E., Rynes, E., Humbert, R., Vierstra, J., Maurano, M.T., Haugen, E., et al., 2012. The accessible chromatin landscape of the human genome. Nature 489, 75–82.

Torre, C., Abnave, P., Tsoumtsa, L.L., Mottola, G., Lepolard, C., Trouplin, V., et al., 2017. *Staphylococcus aureus* promotes smed-PGRP-2/smed-setd8-1 methyltransferase signalling in planarian neoblasts to sensitize anti-bacterial gene responses during Re-infection. EBioMedicine 20, 150–160.

Treer, D., Maex, M., Van Bocxlaer, I., Proost, P., Bossuyt, F., 2018. Divergence of species-specific protein sex pheromone blends in two related, nonhybridizing newts (Salamandridae). Mol. Ecol. 27, 508–519.

True, J.R., Carroll, S.B., 2002. Gene co-option in physiological and morphological evolution. Annu. Rev. Cell Dev. Biol. 18, 53–80.

Ullrich, B., Ushkaryov, Y.A., Südhof, T.C., 1995. Cartography of neurexins: more than 1000 isoforms generated by alternative splicing and expressed in distinct subsets of neurons. Neuron 14, 497–507.

Vandegehuchte, M.B., Lemière, F., Vanhaecke, L., Vanden Berghe, W., Janssen, C.R., 2010. Direct and transgenerational impact on Daphnia magna of chemicals with a known effect on DNA methylation. Comp. Biochem. Physiol., C 151, 278–285.

van de Pavert, S.A., Olivier, B.J., Goverse, G., Vondenhoff, M.F., Greuter, M., Beke, P., et al., 2009. Chemokine CXCL13 is essential for lymph node initiation and is induced by retinoic acid and neuronal stimulation. Nat. Immunol. 10, 1193–1199.

van Spronsen, M., van Battum, E.Y., Kuijpers, M., Vangoor, V.R., Rietman, M.L., Pothof, J., et al., 2013. Developmental and activity-dependent miRNA expression profiling in primary hippocampal neuron cultures. PLoS One 8 (10), e74907.

Vieira, G.C., D'Ávila, M.F., Zanini, R., Deprá, M., da Silva Valente, V.L., 2018. Evolution of DNMT2 in drosophilids: evidence for positive and purifying selection and insights into new protein (pathways) interactions. Genet. Mol. Biol. 41 (Suppl. l), 215–234, 1.

Wada, K., Sakaguchi, H., Jarvis, E.D., Hagiwara, M., 2004. Differential expression of glutamate receptors in avian neural pathways for learned vocalization. J. Comp. Neurol. 476, 44–64.

Wainwright, G., Websters, S.G., Rees, H.H., 1998. Neuropeptide regulation of biosynthesis of the juvenoid, methyl farnesoate, in the edible crab, *Cancer pagurus*. Biochem. J. 334, 651–657.

Wang, E.T., Sandberg, R., Luo, S., Khrebtukova, I., Zhang, L., Mayr, C., et al., 2008. Alternative isoform regulation in human tissue transcriptomes. Nature 456, 470–476.

Wang, Y., Liu, J., Huang, B., Xu, Y.-M., Li, J., Huang, L.-F., et al., 2015. The mechanism of alternative splicing and its regulation. Biomed Rep 3, 152–158.

Watson, F.L., Püttmann-Holgado, R., Thomas, F., Lamar, D.L., Hughes, M., Kondo, M., et al., 2005. Extensive diversity of ig-superfamily proteins in the immune system of insects. Science 309, 1874–1878.

Wayman, G.A., Davare, M., Ando, H., Fortin, D., Varlamova, O., Cheng, H.-Y.M., et al., 2008. An activity-regulated microRNA controls dendritic plasticity by down-regulating p250GAP. Proc. Natl. Acad. Sci. U.S.A. 105, 9093–9098.

Weaver, M., Hogan, B., 2001. Powerful ideas driven by simple tools: lessons from experimental embryology. Nat. Cell Biol. 3, E165–E167.

West, A.E., Greenberg, M.E., 2011. Neuronal activity–regulated gene transcription in synapse development and cognitive function. Cold Spring Harb Perspect Biol 3 (6), a005744, 2011 Jun.

Wheeler, B.M., Heimberg, A.M., Moy, V.N., Sperling, E.A., Holstein, T.W., Heber, S., Peterson, K.J., 2009. The deep evolution of metazoan microRNAs. Evol. Dev. 11, 50–68.

Wibrand, K., Panja, D., Tiron, A., Ofte, M.L., Skaftnesmo, K.-O., Lee, C.S., Pena, J.T., Tuschl, T., Bramham, C.R., 2010. Differential regulation of mature and precursor microRNA expression by NMDA and metabotropic glutamate receptor activation during LTP in the adult dentate gyrus in vivo. Eur. J. Neurosci. 31, 636–645.

Wilburn, D.B., Eddy, S.L., Chouinard, A.J., Arnold, S.J., Feldhoff, R.C., Houck, L.D., 2015. Pheromone isoform composition differentially affects female behaviour in the red-legged salamander, *Plethodon shermani*. Anim. Behav. 100, 1–7.

Wilburn, D.B., Swanson, W.J., 2016. From molecules to mating: rapid evolution and biochemical studies of reproductive proteins. J. Proteomics 135, 12–25.

Will, C.L., Lührmann, R., 2011. Spliceosome structure and function. Cold Spring Harb. Perspect. Biol. 3 (7), a003707.

Wirsig-Wiechmann, C.R., Houck, L.D., Wood, J.M., Feldhoff, P.W., Feldhoff, R.C., 2006. Male pheromone protein components activate female vomeronasal neurons in the salamander *Plethodon shermani*. BMC Neurosci. 7, 26.

Wistow, G., 1993. Lens crystallins: gene recruitment and evolutionary dynamism. Trends Biochem. Sci. 18, 301–306.

Wittkopp, P.J., 2007. Variable gene expression in eukaryotes: a network perspective. J. Exp. Biol. 210, 1567–1575.

Wu, S.C., Zhang, Y., 2010. Active DNA demethylation: many roads lead to Rome. Nat. Rev. Mol. Cell Biol. 11, 607–620.

Wülfing, C., Schuran, F.A., Urban, J., Oehlmann, J., Günther, H.S., 2018. Neural architecture in lymphoid organs: hard-wired antigen presenting cells and neurite networks in antigen entrance areas. Immun. Inflamm. Dis. 6, 354–370.

Xie, J., 2008. Control of alternative pre-mRNA splicing by Ca^{2+} signals. Biochim. Biophys. Acta 1779, 438–452.

Yamanouye, N., Britto, L.R.G., Carneiro, S.M., Markus, R.P., 1997. Control of venom production and secretion by sympathetic outflow in the snake *Bothrops jararaca*. J. Exp. Biol. 200, 2547–2556.

Yang, Y., Yamada, T., Hill, K.K., Hemberg, M., Reddy, N.C., Cho, H.Y., et al., 2016. Chromatin remodeling inactivates activity genes and regulates neural coding. Science 353, 300–305.

Yao, N., Lu, C.L., Zhao, J.J., Xia, H.F., Sun, D.G., Shi, X.Q., et al., 2009. A network of miRNAs expressed in the ovary are regulated by FSH. Front. Biosci. 14, 3239–3245.

Yao, N., Yang, B.Q., Liu, Y., Tan, X.Y., Lu, C.L., Yuan, X.H., Ma, X., 2010. Follicle-stimulating hormone regulation of microRNA expression on progesterone production in cultured rat granulosa cells. Endocrine 38, 158–166.

Yap, E.-L., Greenberg, M.E., 2018. Activity-regulated transcription: bridging the gap between neural activity and behavior. Neuron 100, P330–P348.

Yi, S.V., Goodisman, M.A.D., 2009. Computational approaches for understanding the evolution of DNA methylation in animals. Epigenetics 4, 551–556.

Yu, J.Y., Kanai, M.I., Demir, E., Jefferis, G.S.X.E., Dickson, B.J., 2010. Cellular organization of the neural circuit that drives 1 courtship behavior. Curr. Biol. 20, 1602–1614.

Yun, S., Saijoh, Y., Hirokawa, K.E., Kopinke, D., Murtaugh, L.C., Monuki, E.S., et al., 2009. Lhx2 links the intrinsic and extrinsic factors that control optic cup formation. Dev. Camb. Engl. 136, 3895–3906.

Zemach, A., Zilberman, D., 2010. Evolution of eukaryotic DNA methylation and the pursuit of safer sex. Curr. Biol. 20, R780–R785.

Zhang, H., Hollander, J., Hansson, L.-A., 2017. Bi-directional plasticity: rotifer prey adjust spine length to different predator regimes. Sci. Rep. 7 (2017), 10254.

Zhang, Y.-N., Zhu, X.-Y., Wang, W.-P., Wang, Y., Wang, L., Xu, X.-X., 2016. Reproductive switching analysis of *Daphnia similoides* between sexual female and parthenogenetic female by transcriptome comparison. Sci. Rep. 6 (2016), 34241.

Chapter 4

Cambrian explosion: sudden burst of animal bauplaene and morphological diversification

Chapter outline

Epigenetic Mechanisms of the Cambrian Explosion. https://doi.org/10.1016/B978-0-12-814311-7.00004-4
Copyright © 2020 Elsevier Inc. All rights reserved.

End of the ediacaran fauna

Ediacaran trace fossils are almost monotonous and show very little diversity
(Marshall, 2006): "the overarching pattern of pre-Ediacaran eukaryotes,
including both taxonomically resolved and problematic forms, is one of
minimal morphological diversity and profound evolutionary stasis" (Butterfield, 2007). Earlier the Ediacaran fossil record was even interpreted as artifact
(Gehling, 1991), but most investigators believe in the real existence of Ediacaran biota (Fig. 4.1).

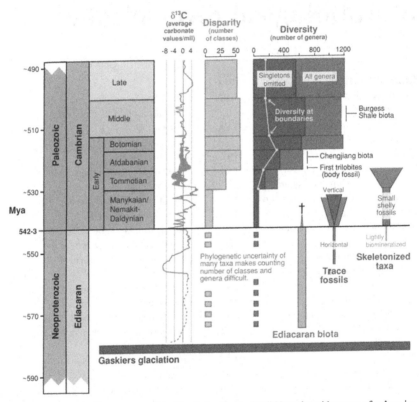

FIGURE 4.1   Complex anatomy of the Cambrian "explosion." Note the wide range of values in
part of the Early Cambrian; this is partly due to geographic variation, but also to variation
measured in Morocco. Disparity from Bowring et al. (1993). Diversity based on tabulation by
Foote (2003) derived from Sepkoski's compendium of marine genera (Sepkoski 1997; 2002); all
taxa found in the interval, as well as those that range through the interval, are counted. Short-term
idiosyncrasies in the rock record can add noise to diversity curves, so to dampen that effect, taxa
found in just one interval can be omitted (singletons omitted). Note that standing diversities were
much lower than the values shown; many of the taxa found in a stratigraphic interval did not
coexist. The boundary crosser curve (M. Foote, personal communication) gives the number of taxa
that must have coexisted at the points shown; however, because traditional stratigraphic boundaries
are based on times of unusual taxonomic turnover, these estimates may underestimate typical
standing diversities. *From Marshall, C.R., 2006. Explaining the cambrian "explosion" of animals.
Annu. Rev. Earth Planet Sci. 34, 355–384.*

There is no confirmed evidence on Ediacaran vertical burrows and what was previously believed to be a meandering trace fossil turned out to be a protist body fossil. Ediacaran deep sea ichnofauna were extremely simple and "ichnodiversity is much lower than it appears from the literature" (Seilacher et al., 2005). Ediacaran ichnospecies (species determined by trace fossils) also do not exhibit foraging behavior or search locomotion patterns. Their locomotion type is observed in Cnidaria, whose general forms of locomotion are slow swimming (medusae) by expelling water jets, but most of time they are passively carried by currents and polyps move by alternating points of contact on substrate.

By the end of the Proterozoic, the Ediacaran fauna also colonized deep sea bottoms (Marshall, 2006), and from this time on, ichnospecies exhibit a behavioral diversification that reached its climax during the Early Cambrian (Seilacher et al., 2005).

Representatives of the Ediacaran biota disappear from the fossil record close to the Cambrian. Their disappearance is characterized as "the first major biotic crisis of macroscopic eukaryotic life" (Buatois et al., 2014).

It was believed that a significant time gap exists between the last Ediacaran representatives and the appearance of *Treptichnus pedum*, which represents the beginning of the Cambrian era. But recent evidence and interpretations seem to prove that a few representatives of the Ediacaran biota persisted into the Ediacaran—Cambrian transition (Buatois et al., 2014). Ediacaran fossils of *Gaojiashania*, *Wutubus* and *Conotubus* and Cloudinia of the family *Cloudinidae* are considered to be transitional Ediacaran—Cambrian metazoans (Smith et al., 2016). Despite the evidence on bilaterian organisms during the end of Ediacaran period (*Kimberella*, etc.), it is admitted that the evolutionary disappearance of the Ediacaran biota and the appearance of arthropod-made scratches are "nearly coincident with the Ediacaran—Cambrian boundary" (Buatois et al., 2014).

An early trace fossil of a segmented bilaterian is that of *Spriggina floundersi* dated during the end of the Ediacaran eon, about 550 Ma. It may be an early trilobite but no special relationship has been established with any extant animal group.

The conventional ediacaran-cambrian boundary

Paleontologists have agreed to recognize a trace fossil first found in rocks of southeastern Newfoundland, Canada, but later throughout Lagerstätten worldwide, as the footprint (trail record) of the animal that set the boundary between the Ediacaran biota and the Cambrian fauna. The burrower ichnospecies is known as *T. pedum* (formerly *Trichophycus pedum*, *Phycodes pedum*, etc.). It is a worm or an arthropod, of several mm long, featuring for the first time complex "turning and stitching" burrows (Fig. 4.2), as a result of repeated probing and withdrawing behavior, indicating that it was a

(A)   **(B)**

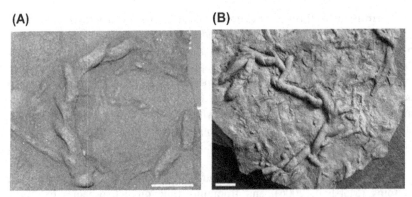

FIGURE 4.2 Behavioral variability of the ichnogenus *Treptichnus*. (A) *Treptichnus pedum*, Klipbak Formation, near Brandkop, South Africa. (B) *Treptichnus rectangularis*, Wiśniówka Sandstone Formation, Wiśniówka Duża Quarry, Holy Cross Mountains, Poland. *From Mángano, M.G., Buatois, L.A., 2016. The Cambrian explosion (Chapter 3). In: M.G. Mángano, L.A. Buatois (Eds.), The Trace-Fossil Record of Major Evolutionary Events. Topics in Geobiology, vol. 39, Springer, Dordrecht, pp. 77 (73–126).*

bilaterian organism that had evolved a considerable centralization of the nervous system.

Being a soft-bodied metazoan, it left no body fossils, but its trace patterns indicating forward and upward probes were a dramatic evolutionary novelty suggesting it has been a bilaterian with a centralized nervous system. Among the extant animals, the unsegmented marine worms of the phylum *Priapulida*, whose fossils are dated from Middle Cambrian, are believed to be their crown group: "treptichnid burrow systems were most probably produced by priapulid worms or by worms that used the same locomotory mechanisms as the recent priapulid" (Vannier et al., 2010). It is interesting the fact that the LAD (last appearance datum) of the tubular Ediacaran forms, *Conotubus and Wutubus*, coincide with the FAD (first appearance datum) of *T. pedum* (Smith et al., 2016).

Since the middle of the 20th century, *T. pedum* complex trace fossils were found in many sites around the world. Its behavior is "iconic of the Cambrian explosion" (Buatois et al., 2018). While treptichnids are found below the Ediacaran–Cambrian border, they are absent in Ediacaran rocks (Buatois, 2018).

Using as landmark the first appearance in rocks of the trace fossil *T. pedum*, palaeontologists have, with relatively great precision, determined the Ediacaran–Cambrian border at 542 million years ago (Fig. 4.3) and quantitative evidence has shown that the modern-style Cambrian fauna during the first 20 million years was "followed by broad-scale evolutionary stasis throughout the remainder of the Cambrian" (Paterson et al., 2019).

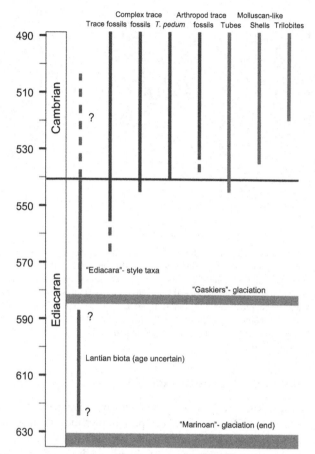

FIGURE 4.3 A broad outline of important aspects of the fossil record from the Ediacaran to Cambrian. Dating from, e.g., Bowring et al., 2007. *Treptichnus pedum* is the trace fossil that formally marks the base of the Cambrian. *From Budd, G.E., 2013. At the origin of animals: the revolutionary cambrian fossil record. Curr. Genomics 14, 344–354.*

The "first" Cambrian animal is still waiting to be firmly confirmed. It is assigned the role of Ediacaran–Cambrian watershed because, as opposed to any Ediacaran fossil traces, *T. pedum* fossil traces in seafloor indicate three-dimensional burrowing, not only back and forth in the horizontal plane but also vertically, up, and down in search of food: "These branching burrow systems record the appearance of the first complex behaviors, representing the prelude of the dramatic increase in complexity that took place subsequently as a result of the Cambrian explosion" (Mángano and Buatois, 2016).

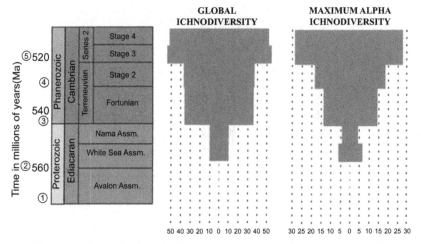

FIGURE 4.4 Summary diagram of changes in global and alpha ichnodiversity during the Ediacaran–Cambrian transition (after Mángano and Buatois, 2014; Buatois et al., 2014). 1 = appearance of the Ediacaran biota; 2 = first uncontroversial evidence of bilaterian trace fossils; 3 = major diversification of trace-fossil bauplans; 4 = onset of vertical bioturbation and coupling of benthos and plankton; 5 = earliest fossil Lagerstätte (Chengjiang) and Cambrian explosion according to body fossils. In contrast to fossil Lagerstätte, the trace-fossil record is continuous through the critical Ediacaran–Cambrian interval. Assm. = assemblage. *From Mángano and Buatois (2016).*

On the reality of the Cambrian explosion

The Cambrian explosion is a sudden evolutionary "Simpsonian" event that occurred during an evolutionarily short period of time starting about 542 million years ago (Fig. 4.4). It is seen as an outburst essentially of the bilaterians that evolved the mechanism of terminal addition of body segments (Jacobs et al., 2005). Most investigators believe in the reality of the Cambrian explosion (Eldredge and Gould, 1972; Conway Morris, 2000; Nichols et al., 2006). Some recent evidence dates the Ediacaran–Cambrian boundary about 4 million years earlier than assumed, to 538.6–538.8 Ma. The time interval between the extinction of the Ediacaran biota (except sponges and possibly cnidarians) and the beginning of the Cambrian radiation was an extremely short period of time, of only ~400,000 years (Linnemann et al., 2019).

Some authors, however, believe that the divergence of metazoan clades occurred much earlier in the Earth's history, and the Cambrian explosion is rather part of the Darwinian gradual process of evolution (Fortey et al., 1996; Ayala et al., 1998; Lieberman, 2003). But there are two serious arguments against their conjectural conclusion. Firstly, that their estimates rely on molecular clock based on gene substitutions, which are generally uncertain and erratic because, even within clades, the rates of change are different for different genes and vary over time (Ayala, 1997; Valentine et al., 1999). This is incontestably exposed by the fact that molecular clock estimates of the time of divergence of metazoan clades by different authors led to values that vary widely, ranging from more than 1500 Ma to 670 Ma (Ayala et al., 1998). So, e.g., Runnegar estimated the time of divergence of animal phyla (protostome–deuterostome) at 700 million years (Runnegar, 1986), whereas Wray et al. (1996) believe that the divergence began first in mid-Proterozoic about 1200 million years ago, and Wang et al. (1999) at 1000 Ma, i.e., pre-dating Cambrian explosion by 400–600 million years.

Secondly, that no fossil record or other empirical evidence exist to substantiate or corroborate their estimates on the proterozoic origin of metazoans and protostome–deuterostome divergence.

Deniers of the Cambrian explosion have also conjectured that Cambrian explosion may be a paleontological artifact, a result of nonfossilizability of soft-bodied, pre-Cambrian metazoans that evolved long before, but unlike Cambrian fauna did not evolve skeletons to leave body fossils. The hypothesis is rejected from the paleontological evidence of fossils of soft-bodied Ediacaran and Cambrian metazoans, as well as the fact that microscopic, nonmetazoan fossils of soft-bodied unicellulars of that age abound; most of the known Cambrian fossils are from soft-bodied metazoans (Gould, 1989; Conway Morris, 1998) and skeletization of metazoans is not a universal feature of the Cambrian explosion. Moreover, if a significant radiation of metazoans would have occurred in the Precambrian era, it would be expected that hard-bodied (with skeleton or exoskeleton) animals would have evolved at that time as well. Thus, the absence of fossilized animals in the pre-Cambrian paleontological record also goes against the hypothesis of the gradual evolution and diversification radiation of metazoans during the pre-Cambrian. Besides, the Cambrian fauna is characterized not only by appearance of hard-bodied animals but also, above all, by a diversification of morphology and numerous body plans, including 30 modern phyla.

Moreover, the "explosive" evolution of the Cambrian is also observed in the trace fossils that reflect rapid evolution of the complexity of animal behavior, its neurobiological underpinnings, and acceleration of the centralization of the nervous system during the Cambrian radiation.

Now biologists, almost by consensus, believe in the reality of the Cambrian explosion and admit that "there is nothing like the Cambrian until the Cambrian..." (Knoll, 2004) and also "... after the Cambrian" (Mángano and Buatois, 2016).

The generally accepted fossil record of animals begins at the base of the Cambrian, around 542 million years ago, with the first finding of animal-like trace fossils (Budd, 2013). Since the sponges and cnidarians entered "dead ends" of evolution, the Cambrian explosion is essentially a rapid diversification of bilaterians alone. It is characterized by a "top-down" pattern, that is, the phyla diversity preceded the evolution of the classes and orders within the phyla as opposed to the latter evolution that produced only one phylum, that of bryozoans, but more classes, orders, and genera (Budd, 2013).

Low cambrian trace fossils

Trace fossils of the Early Cambrian are distinctive and clearly more complex. *T. pedum* looping and spiraling trails suggest that the animal had a more complex nervous system. The burrowing movements through sediments were feeding-motivated. In clear distinction from exclusively horizontal trace fossils left by Ediacaran surface moving/scratching forms, early Cambrian metazoans added vertical movements to their foraging behavior, implying that the pioneering proto-Cambrian bilaterians had evolved three-dimensional perception of their foraging space. Hence, *T. pedum* seems to have been a soft-bodied bilaterian "with a well-developed nervous system, anterior-posterior asymmetry, and a one-way gut" (Schopf, 1992) and, in all likelihood, specialized organs for locomotion. Another type of trace fossil left by small animals at the time was characterized by meandering trails and the reversion of the direction of the movement, a behavior that requires very few "instructions" (Schopf, 1992).

During the Early Cambrian appear the first problematic fossil groups, phyla, and subphyla with distinctive skeletons and Bauplaene (Valentine, 1994). The first confirmed body fossil may be that of *Vittatusivermis annularius* gen. et sp. nov., found in phosphoritic rocks of lower Cambrian in South China; the worm-like organism is a bilaterian about 1 cm wide and more than 26 cm in preserved length (Zhang et al., 2017).

Among early Cambrian bilaterians are fossils of *Oldhamia*, the only distinctive ichnospecies identified so far in shallow-marine environments. Its behavioral diversity and the disparity of life styles initially were very low, but later during the Early Cambrian, *Oldhamia* displays a remarkable behavioral diversification (Marshall, 2006) (Fig. 4.5).

Evolution of bilateral symmetry

All metazoans, except Cnidaria and Ctenophora, and nerveless Porifera and Placozoa, belong to the supergroup Bilateria. The origin and nature of the Urbilateria is not resolved and continues to be subject to controversy. What can be stated with a high degree of probability is what can be inferred from the appearance of the first generally accepted Precambrian fossils of the bilaterian organism, *Kimberella*, about 558 million years ago. The identification of *Kimberella* as a protostome organism suggests that the division protostome—deuterostome in bilaterians (nephrozoans) occurred before the Cambrian, whereas the cnidarian—bilaterian split may have occurred about 570—580 Mya (Erwin and Davidson, 2002).

While nerveless sponges and placozoans have no strictly determined body symmetry, cnidarians evolved radial symmetry. The transition from the

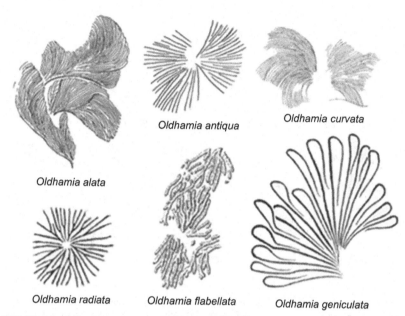

Oldhamia antiqua

Oldhamia curvata

Oldhamia alata

Oldhamia radiata

Oldhamia flabellata

Oldhamia geniculata

FIGURE 4.5   Behavioral variability of ichnogenus *Oldhamia* as illustrated by its high diversity at ichnospecies level in the early Cambrian. *Modified from Seilacher, A., Buatois, L.A., Mángano, M.G., 2005. Trace fossils in the Ediacaran-Cambrian transition: behavioral diversification, ecological turnover and environmental shift. Palaeogeogr. Palaeoclimatol. Palaeoecol. 227, 323—356; from Mángano, M.G., Buatois, L.A., 2016. The Cambrian explosion (Chapter 3). In: M.G. Mángano, L.A. Buatois (Eds.), The Trace-Fossil Record of Major Evolutionary Events. Topics in Geobiology, vol. 39, Springer, Dordrecht, pp. 77 (73—126).*

cnidarian radial symmetry to bilaterality (animals with right and left sides) is one of the major transitions in eumetazoan history (Fig. 4.6). Trace fossils show that some bilaterian organisms appear by the end of the Ediacaran eon. Bilaterality was of critical importance for the Cambrian explosion. It not only enabled easier streamlined intentional locomotion that increased metazoans' ability to faster approach sources of nutrition, chase the prey, and escape predators but also facilitated the evolution of more complex and diversified animals by multiplying the amount of positional information (Genikhovich and Technau, 2017). The advent of bilaterality in animals coincided with (and probably is related to) the emergence of the centralization of the nervous system, as well as with the advent of triploblasty (the evolution of mesoderm, the third embryonic layer), which implies the evolution of the ability of the directed movement of particular groups of cells (Meinhardt, 2006).

Nothing certain can be said about the mechanism(s) that made possible the major transition from the radial to bilateral symmetry. However, a conspicuous fact related to this transition is the coincidence of the evolution of the

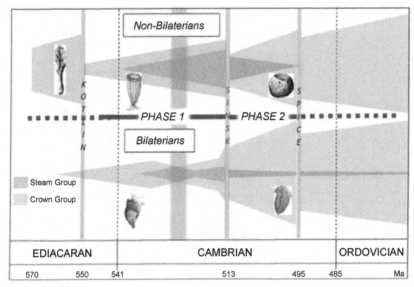

FIGURE 4.6   Schematic of hypothesized nonbilaterian (total group Porifera, Cnidaria, and Ctenophora) and Bilaterian diversification during the Ediacaran–Cambrian metazoan radiation, showing the fossil record of probable earliest metazoans (shown by a rangeomorph reconstruction), the Kotlin crisis, followed by two phases of Cambrian Explosion, separated by the Sinsk event extinction (with a possible expanded interval of anoxia during Phase 1) and extending to the Ordovician Radiation through the SPICE extinction. Nonbilaterian stem group example is a stem group archaeocyath sponge; crown group is a crown group demosponge. Bilaterian stem group is shown by a tommotiid; crown group by a trilobite. *From Zhuravlev, A.Y., Wood, R.A., 2018. The two phases of the Cambrian Explosion. Sci. Rep. 8, 16656.*

centralized bilobed brain, on the one hand, and the advent of bilateral symmetry, the advent of the ventrolateral axis, specialized sense organs and organ systems (digestive, circulatory, and excretory), on the other. The evolution of the CNS facilitated the intentional locomotion of bilaterians toward the sources of nutrition to chase the prey, escape predators, etc. From an evolutionary viewpoint, it is also worth mentioning that planula larvae of a radially symmetric organism, such as the cnidarian sea anemone, *Nematostella vectensis*, develops a concentration of neurons in the anterior (oral) part of the body (Watanabe et al., 2014) and bilateral body, in stark contrast with the mature form of the cnidarian that has a diffuse nervous system and radial symmetry.

The coincidence of the evolution with the bilaterality, the evolution of the brain, emergence of sense organs and organ systems may be a contingent occurrence, but there is no visible reason to rule out a causal relationship in their simultaneous emergence.

Evolution of body segmentation

Some metazoan groups in the early Cambrian evolved armor plates, the dermal sclerites (Bengtson, 2001). But development of hard exoskeletons was an obstacle to the lateral movements of early eumetazoans and served as a selective pressure for evolving joints dividing the body in repetitive segments, which enabled and facilitated lateral movements of the early armored animals. For the first time during the Cambrian, bilaterians evolved the body segmentation that is the ability to produce repetitive body segments along the anterior–posterior axis. The evolutionary advantage of segmentation is illustrated by the fact that three of the most speciose phyla in the animal kingdom arthropods, annelids, and chordates have segmented bodies (Fig. 4.7). As an alternative driving force for evolution of segmentation, Chipman considers regionalization of the nervous system whose "reiterated ganglia" served as "precursors to full segmentation" (Chipman, 2009). Segmentation has repeatedly evolved in bilaterians (Graham et al., 2014) indicating its functional selective advantages in regard to locomotion. Body segmentation in bilaterians occurs in the form of homomerization (addition of similar parts along the anterior–posterior body axis) or heteromerization. Controversies arise on whether the evolution of segmentation in bilaterians is result of "amplification" or "parcellation" of the bilaterian body, whether the unsegmented

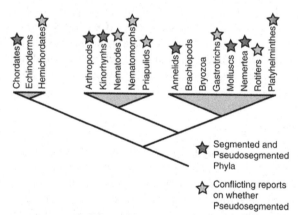

FIGURE 4.7 Phylogenetic relationship among segmented and unsegmented phyla. Segmentation is no longer a rare characteristic if both segmented and pseudosegmented phyla are considered (red stars mark groups identified as segmented or pseudosegmented in several papers, orange stars mark groups identified as segmented or pseudosegmented in one paper and unsegmented in another). Here, "pseudosegmented" is meant solely to distinguish traditionally segmented chordates, arthropods and annelids from other phyla with repetition of units with anterior–posterior polarity along the anterior–posterior axis. It does not necessarily mean that there is a biological distinction between these groups based on their repeated units. *From Hannibal, R.L., Patel N.H., 2013. What is a segment? EvoDevo 4, 35.*

ancestor experienced addition of repetitive segments or whether its body was divided in repetitive parts.

It is a complex question to tell whether the development can reveal us anything about the evolution of segmentation, but it is worth mentioning that during the velvet worm (onychophoran) development the ventral nerve tract grows posteriorly from clusters of neuroblasts at an early stage and the nerves go out of the segmentally organized nervous system to support limb formation before the limb starts developing, a process that Budd characterizes as "The imposition of segmental structure onto an essentially unsegmented body region (the epidermal ectoderm) from the nervous system" (Budd, 2001), alluding to existence of a cause−effect relationship between the nervous system and the segmentation.

Indeed, based on earlier observations and their own observations on development of the crustacean *Parhyale hawaiensis,* Hannibal et al. confirmed conclusions of earlier investigators that "a segmented nervous system evolved prior to a segmented epidermis and this evolutionary scenario could be reflected in developmental mechanisms of extant animals" (Hannibal et al., 2012). Earlier also was noticed that the incipient nervous system in vertebrate embryos determines "the development of paired elements such as the somites that presage the vertebrae, and paired organ rudiments such as left and right limb buds and the primordia of the gonads, kidney, lung and heart" (Hall, 1998a). Similarly, it is observed that innervation in limb buds precedes the development of limbs in tetrapods (Berggren et al., 1999) and in the development of presomitic segments and somites as primitive segments of vertebrates (Cossu and Borello, 1999; Fan and Tessier-Lavigne, 1994). Thus, formation of presomitic segments demonstrates that existence of a segmentation clock in the neural tube/notochord axis (Cabej, 2018a) regulates rhythmic expression of PSM (presomitic mesoderm) segments.

Cambrian fossils

Panarthropoda is a superphylum that comprises phyla Arthropoda, Onychophora, and Tardigrada, with the latest group's position still uncertain. *Panarthropoda* are widely represented in the Cambrian fauna. A lobopod may have been the ancestor of tardigrades and a lobopod with appendage articulations like the Lower Cambrian, *Fuxianhuia protensa* isp. Hou, 1987 (Fig. 4.8) may have been ancestor of euarthropods and onychophorans (Xianguang and Bergstrom, 1997).

There is a substantial fossil record of arthropods and the total number of the extant arthropod species, most of them insects, amounts to several millions. The earliest euarthropod trace fossil records are dated around 537 Ma (an earlier origin of euarthropods seems to be disproven by the lack of such trace fossils in Ediacaran Lagerstätten), but many believe they cannot be older

(A)   (B)

FIGURE 4.8   Fossil of the arthropod *Fuxianhuia protensa* Hou, 1987. (A) Fossil specimen with articulated limbs from Maotianshan Hill, Chengjiang County, Yunnan Province, China. (B) Reconstruction of the fossil specimen of *F. protensa* Hou, 1987. *From Xianguang, H., Bergstrom, J., 1997. Arthropods of the lower Cambrian Chengjiang fauna, southwest China. Foss. Strata 45, 1–116.*

than 550 Ma (Daley et al., 2018). Crown group arthropods appear at the base of the Atdabanian, ~ 530–524 Ma, and the earliest fossil of the group may be *Rusophycus*, commonly *associated with trilobites*, which appears before trilobites, earlier at the Tommotian (Budd and Jensen, 2017).

The fossil record has shown that early Cambrian radiodontan arthropods possessed large compound eyes (Cong et al., 2016) (Fig. 4.8).

Panarthropoda are characterized by appearance of the ventral nerve cord and the central nervous system in *Euarthropoda, Tardigrada*, and *Chengjiangocaris kunmingensis*. They evolved segmental ganglia, whereas the latter developed regularly spaced nerve roots similarly to the VNC of Onychophora, whose postcephalic CNS is lateralized and lacks segmental ganglia (Yang et al. (2016) (Fig. 4.9).

†, fossil taxa; ?, uncertain character polarity within total-group Euarthropoda; asn, anterior segmental nerve; cn, longitudinal connectives; co, commissure; dln, dorsolateral longitudinal nerve; ico, interpedal median commissure; irc, incomplete ring commissure; pn, peripheral nerve; psn, posterior segmental nerve; rc, ring commissure. Reconstruction of VNC in Onychophora adapted from Dewel et al. (1993). Distribution of serotonin in the trunk of *Metaperipatus blainvillei* (Onychophora, Peripatopsidae): implications for the evolution of the nervous system in Arthropoda. J. Comp. Neurol. 507(2), 1196–1208.

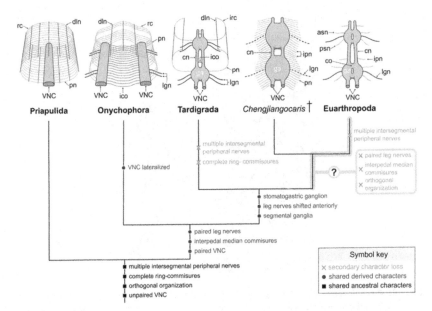

FIGURE 4.9 Simplified cladogram showing the evolution of the postcephalic CNS in *Panarthropoda*. The topology supports a single origin for the condensed ganglia (ga) in the VNC in a clade including Tardigrada and Euarthropoda; note that the presence of multiple intersegmental peripheral nerves (ipn) in *Chengjiangocaris kunmingensis* represents an ancestral condition. Given the morphological similarity between peripheral and leg nerve roots, the presence of a single pair of leg nerves (lgn) in *C. kunmingensis* is hypothetical (*dashed lines*) and based on the condition observed in crown-group Euarthropoda. *From Yang, J., Ortega-Hernández, J., Butterfield, N.J., Liu, Y., Boyan, G.S., Hou, J.-B. et al., 2016. Fuxianhuiid ventral nerve cord and early nervous system evolution in Panarthropoda. Proc. Natl. Acad. Sci. U.S.A. 113: 2988–2993.*

About 540 Ma, Cambrian panarthropods evolved four main brain types, and a great diversity of complex behaviors (Fig. 4.10), but they evolved no novel ground patterns ever since (Strausfeld et al., 2016). The central nervous system of the Cambrian arthropods is largely conserved in extant arthropoda.

Trilobites are derived arthropods and they may have appeared first at the Ediacaran-Cambrian boundary, but the earliest trilobite body fossils are dated 521 Ma (earlier soft-bodied trilobites left no traces), from Siberia (*Profallotaspis jakutensis* and *Profallotaspis tyusserica*), Morocco (*Hupetina antiqua*), Spain (*Lunagraulos tamamensis*), and Laurentia (*Fritzaspis generalis*) followed, within a few million years, by finds into other areas of the earth (Daley et al., 2018). Another trilobite-like arthropod, *Arthroaspis bergstroemi*, is discovered in Sirius Passet, Groenland (Stein et al., 2013). The first trilobite radiation event occurred between 520 and 513 Ma (Lin et al., 2006) (Fig. 4.11) and the group of trilobites existed 20–70 million years before they were extinct (Lieberman, 2008). Trilobite-like fossils like genus Parvancorina

(A)   **(B)**

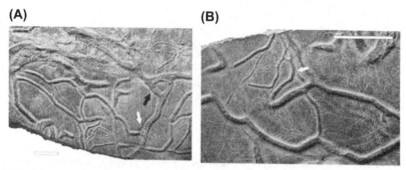

FIGURE 4.10   Trace fossils associated with nonbiomineralized carapaces in Burgess Shale—type deposits. (A) The arthropod *Arthroaspis bergstroemi* showing regular polygonal networks displaying true branching (*black arrows*) and secondary successive branching (*white arrow*), Sirius Passet, Greenland. (B) Close-up of (A) showing annulated structures (*arrow*) and delicate, narrow-caliber, filament-like structures mostly confined to areas among network branches). *From Mángano, M.G., Buatois, L.A., 2016. The Cambrian explosion (Chapter 3). In: M.G. Mángano, L.A. Buatois (Eds.), The Trace-Fossil Record of Major Evolutionary Events. Topics in Geobiology, vol. 39, Springer, Dordrecht, pp. 77 (73—126).*

FIGURE 4.11   *Kootenia* sp, a typical trilobite from the Middle Cambrian of Greenland, a product of the enormous radiation of trilobites that took place around 520 Ma. The body is approximately 87 mm long. *From Budd, G.E., 2013. At the origin of animals: the revolutionary cambrian fossil record. Curr. Genomics 14, 344—354.*

(Lin et al., 2006), are found toward the end of Ediacaran in a few Lagerstätten in Burgess Shale, Australia, and South China.

The head of trilobites evolved of various combinations of body segments. Recently fossil evidence shows that the extinct trilobite *Schmidtiellus reetae*

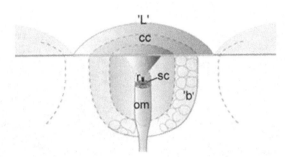

FIGURE 4.12 Schematic drawing of the visual unit of *S. reetae*. *b*, basket; *cc*, crystalline cone; *L*, lens; *om*, ommatidium; *p*, pigment screen; *r*, rhabdom; *sc*, sensory (receptor) cells. (Scale bar: 200 μm). *From Schoenemann, B., Pärnaste, H., Clarkson, E.N.K., 2017. Structure and function of a compound eye, more than half a billion years old. Proc. Natl. Acad. Sci. U.S.A. 114, 13489—13494.*

Bergström, 1973, possessed sophisticated eyes of apposition type, like those of extant bees or dragonflies (Schoenemann et al., 2017) (Fig. 4.12).

Molluscs. The fossil record of the lower Cambrian includes clades of the phylum *Mollusca*, comprising seven extant classes. The crown group *Mollusca* had already evolved in the beginning of the Tommotian, more than 530 Ma, but no consensus exists about their stem group lineage (Budd and Jensen, 2017). Besides the controversial Ediacaran *Kimberella*, among the clades that left fossils in the early Cambrian are gastropods, bivalves, and rostroconchs, as well as helcionellid molluscs, which appear as the earliest gastropods (Landing et al., 2002). Many believe stem group molluscs evolved in the terminal Ediacaran (∼542 Ma) but fossils of mollusc species, along the brachiopods, appear first in the Fortunian ∼537 Ma (Zhuravlev and Wood, 2018). Helcionellids, the small mollusc fossils, commonly seen as molluscs, appear first at the very beginning of the Cambrian over 540 Ma (Steiner et al., 2007) until 530 Ma.

Another group of sclerite-bearing metazoans of the lower Cambrian are halkieriids, a crown group molluscs, among which the better studied, *Halkieria evangelista*, from the Sirius Passet fauna of North Greenland (Conway Morris, 1998), together with *Wiwaxia* and *Odontogriphus* belong to the stem group of the superphylum Lophotrochozoa (Butterfield, 2006).

Fossils of phylum Brachiopoda appear in the terminal Ediacaran (∼542 Ma). This phylum is widely represented in the Cambrian fossil record. Its representatives are among the first skeletal organisms of the Lower Cambrian. Brachiopod fossils are found around the world and in a large number of centers of diversification (Ushatinskaya, 2008). The evolutionary relationship of this group with the extinct lower Cambrian chancelloriids and the extant Lophotrochozoan phyla - molluscs, annelids and brachiopods is not resolved.

(A)   **(B)**

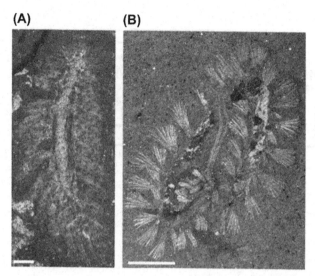

FIGURE 4.13   Cambrian fossil polychaetes and their relationships. (A) *Phragmochaeta cani-cularis*, Geological Museum of Copenhagen MGUH 30888, scale bar 1.5 mm; (B) *Burgessochaeta setigera*, Smithsonian Museum of Natural History 198705, scale bar 2 mm. *From Parry, L., 2014. Fossil focus: annelids, Palaeontol. 4, 1—8.*

Annelids. Annelids are another large phylum of the superphylum Lopho-trochozoans comprising more than 9000 species (Zhang, 2011). The oldest fossils of the phylum Annelida belong to the class Polychaeta found in early Cambrian mudstones (520 million years old) from the Sirius Passet deposit of North Greenland (Parry, 2014) (Fig. 4.13). *Phragmochaeta canicularis* gen. et sp. nov. is an early Atdabanian (~530 Ma) fossil, composed of about 20 segments, from the Lower Cambrian Sirius Passet (Greenland) that is inter-preted as a polychaete annelid (Conway Morris and Peel, 2008). Another likely late Early Cambrian annelid fossil is that of *Myoscolex ateles* Glaessner, 1979, found in Australia (Dzik, 2004).

Hemichordates: Hemichordate fossils abound beginning with the lower Cambrian (Bengtson, 2004). In hemichordates, the branchial slits that earlier were used for filter feeding evolved into pharyngeal slits and into gill slits in fish. A middle Cambrian Burgess Shale fossil, *Oesia disjuncta*, previously compared to annelids and tunicates, now is considered a suspension feeder hemichordate (enteropneust) that dwelled inside tubes of the green algae *Margarita* (Nanglu et al., 2016).

The Yunnanozoan fossils found in the Chengjiang fossil Lagerstätte, South China, are interpreted as the earliest hemichordate fossils and are considered as a link between the invertebrates and vertebrates Shu et al., 1996a,b; Chen et al. (1999).

Phylum *Chordata* comprises three subphyla: subphylum *Cephalochordata* (*Acrania* or lancelets), subphylum *Tunicata* (earlier *Urochordata*) and

(A)

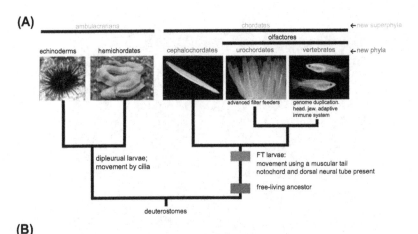

(B)

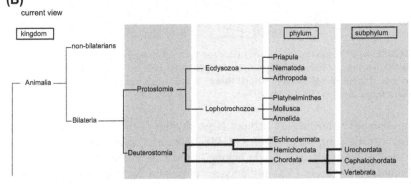

FIGURE 4.14 Phylogenic relationships of deuterostomes and evolution of chordates. (A) Schematic representation of deuterostome groups and the evolution of chordates. Representative developmental events associated with the evolution of chordates are included. (B) A traditional view. FT, *fish-like or tadpole-like. From Satoh, N., Rokhsar, D., Nishikawa, T., 2014. Chordate evolution and the three-phylum system. Proc. Biol. Sci. 281, 20141729.*

subphylum *Vertebrata* (Fig. 4.14). Urochordata and Vertebrata are sister groups, but urochordates have lost segmentation (Nielsen, 2012, p. 348). Their common Bauplan is characterized by

1. segmented body with paired articulated legs,
2. the presence of the notochord, formed from the roof of the archenteron,
3. the neural tube, formed from the ectoderm in contact with the notochord,
4. longitudinal muscles running along the notochord (Nielsen, 2012b, p. 349).

A crucial event in the evolution of the subphylum Vertebrata compared to its sister groups, cephalochordates and tunicates, is the emergence of the neural crest, which was essential for the exceptional increase in the structural and functional complexity of vertebrates (Fig. 4.15).

amphioxus   vertebrates

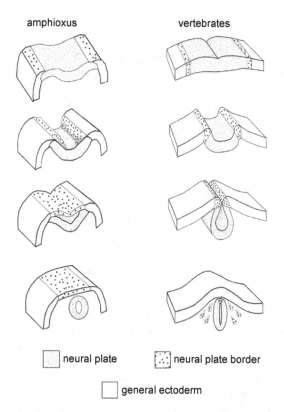

☐ neural plate   ⬚ neural plate border

☐ general ectoderm

FIGURE 4.15 Neurulation in amphioxus and vertebrates. Top: at the late gastrula stage both amphioxus (Cephalochordata) and vertebrates have a neural plate with a neural plate border region. Second from top: at the early neurula stage, in amphioxus, the neural plate border region detaches from the edges of the neural plate and moves over it by lamellipodia. By contrast, in vertebrates, the neural plate border region remains attached to the neural plate as it rounds up. Third from top: at the late neurula stage, in amphioxus, the free edges of the neural plate border region fuse in the dorsal midline, and the neural plate begins to round up underneath the dorsal ectoderm. In vertebrates at a comparable stage, the neural tube has completed rounding up. Bottom: In amphioxus, the neural plate rounds up completely and detaches from the ectoderm. In vertebrates, the neural tube detaches from the ectoderm, and the neural plate border region gives rise to neural crest cells that migrate below the ectoderm and give rise to such structures as pigment cells, cells of the adrenal medulla, parts of cranial ganglia. *From Holland, L.Z., 2015. The origin and evolution of chordate nervous systems. Phil. Trans. R. Soc. B 370, 20150048.*

The evolution of chordates is an integral part of the Cambrian explosion although Cambrian chordate fossils are scanty (Chen and Li, 1997). The earliest Cambrian fossil identified and widely accepted as a basal chordate is genus Pikaia represented by a single species *Pikaia gracilens* (Conway Morris and Whittington, 1979) dated to the Middle Cambrian (Fig. 4.16).

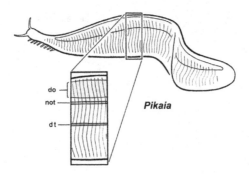

FIGURE 4.16 *Pikaia gracilens*, as reconstructed by Conway Morris and Caron [1]. The head bears a pair of tentacles, probably sensory in nature, and paired rows of ventrolateral projections that may be gills. Not shown: the expanded anterior (pharyngeal) region of the digestive tract, and the dorsal shield-like structure, the anterior dorsal unit, that lies above it. The boxed detail shows the main axial features: the dorsal organ (do), and the putative notochord (not) and digestive tract (dt). The size range among specimens is 1.5−6 cm, which makes this animal very close in size to the adult stage of modern lancelets (amphioxus). *From Lacalli, T., 2012. The Middle Cambrian fossil Pikaia and the evolution of chordate swimming. EvoDevo 3, 12.*

Cambrian fossils of *P. gracilens* are described as basal to chordates. It had a notochord and about 100 myomeres, resembling chordate myomeres, and notochord. Pikaia was not a fast swimmer because its monomers contained only slow twitch muscle fibers. It is considered as "the most stem-ward of the chordates with links to the phylogenetically controversial yunnanozoans" (Conway Morris and Caron, 2012).

Later, another chordate fossil, *Cathaymyrus diadexus*, about 10 million years older than Pikaia, was discovered in South China Lagerstätte (Shu et al., 1996a,b). Chordate fossils and even agnathan fishes are part of the Chengjiang Cambrian fauna in South China found in deposits of 545−490 Ma, but chordate fossils evolved as early as ∼555, if not earlier (Shu et al., 1999).

Commonly tunicates are considered to be the closest relatives to vertebrates. They conserve the notochord only during larval development but lose it as adults, after metamorphosis. All the tunicate fossils described as such so far are controversial. The only one that is widely accepted and bears a striking resemblance to modern ascidians is a tunicate species, identified as *Shankouclava shankouense* from the Lower Cambrian Maotianshan Shale, South China, dated abut 520 million years ago (Chen et al., 2003) (Fig. 4.17).

Lower Cambrian fossils of *Haikouella lanceolata* and a few other similar fossils of Chengjiang Lagerstätte of Yunnan province (South China) are identified as jawless chordate fish (agnathans) related to the extant hagfish and lampreys. Of the same Chengjiang origin are fossils of two other chordate taxa *Haikouichthys* and *Zhangjianichthys* (Conway Morris, 2008).

Based on the similarities between the ciliary bands of the echinoderm and enteropneust larvae and the neural tube of amphioxus as well as in the patterns

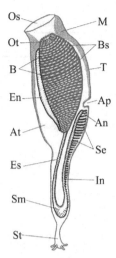

FIGURE 4.17   Lower Cambrian tunicate *Shankouclava* reconstructed. *Abbreviations*: An, anus; Ap, possible atrial pore; At, atrium; B, branchial bars; Bs, branchial slits; En, endostyle; Es, esophagus; In, intestine; M, mantle; Os, oral siphon; Ot, oral tentacle; Se, body segments; Sm, stomach; St, stalk. *From Chen, J.-Y., Huang, D.-Y., Peng, Q.-Q., Chi, H.-M., Wang, X.-Q., Feng, M., 2003. The first tunicate from the early cambrian of south China. Proc. Natl. Acad. Sci. U.S.A. 100, 8314—8318.*

of gene expression in the anterior protostomian CNS and in the brain of vertebrate embryos, biologists (Arendt et al., 2008; Nielsen, 2013d) hypothesize that an inversion of the nerve cord and dorso-ventral organization occurred in the chordate CNS, which evolved via the circumoral ciliary band of a dipleurula, a hypothetical ancestral form of bilaterian echinoderms and chordates. Hence, the dorsal side of chordates is thought to be homologous to the ventral side of protostomes and enteropneusts. Similarly, the vertebrate eyes are considered to be homologous to the ciliary frontal eye of amphioxus (Nielsen, 2012c).

Transition to bilateria in extant paradigms

Cambrian explosion is essentially eruptive evolution of the bilateria, the greatest basic taxon, excluding Cnidaria, Porifera and Placozoa. In the following is provided a brief description of *Xenacoelomorpha*, and Planaria, two of the simplest groups of extant animals believed to have participated in the Cambrian explosion.

Xenacoelomorpha

Xenacoelomorpha are a basal bilaterian phylum of mainly marine worms consisting of two subphyla, Xenoturbellida, with only two species,

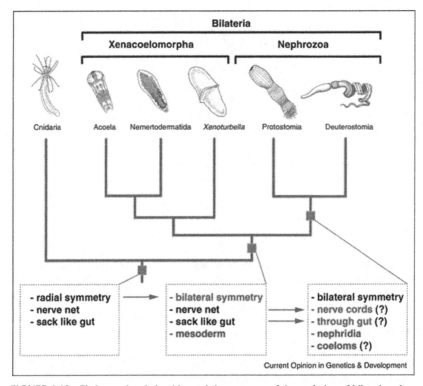

FIGURE 4.18 Phylogenetic relationships and the sequence of the evolution of bilaterian characters. Significant organ systems and their major transitions (*red [gray in print version] arrows*) and novelties (*red [gray in print version]*) mapped on the phylogeny. *From Hejnol, A., Pang, K., 2016. Xenacoelomorpha's significance for understanding bilaterian evolution. Curr. Opin. Genet. Dev. 39, 48–54.*

Xenoturbella bocki and *Xenoturbella westbladi* (4 more species are discovered in 2016) and *Acoelomorpha* with the latter comprising Acoela (~400 species) and Nemertodermatida (10 species), until recently considered to be separate clades (Hejnol and Wanninger, 2015; Hejnol, 2016). Now, many biologists consider the taxon as "the first offshoot of the bilaterians" (Martinez et al., 2017) (Fig. 4.18), although their position is not consensually determined.

Morphology

Xenoturbellida are small marine organisms, up to 4 cm in length (Brusca et al., 2016) (Fig. 4.19). They lack reproductive organs or clearly organized gonads, even though they produce eggs. Sperms and ova are found in various parts of the adult worm (Nakano, 2015). They lack a body cavity but have a midventral mouth and a sack-like digestive organ but lack an anus. They have neither

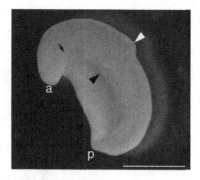

FIGURE 4.19 *Xenoturbella* morphology and collections. A external morphology of *Xenoturbella bocki*. Side furrows (*black arrow*) are present on the lateral sides from the anterior tip (A), but these do not reach the posterior end (p). The mouth (*black arrowhead*) is situated on the ventral side anterior to the circumferential furrow (*white arrowhead*). Scale bar: 1 cm. *From Nakano, H., 2015. What is Xenoturbella? Zool. Lett. 1, 22.*

circulatory nor excretory systems, as well as protonephridia and eliminate waste nitrogen and carbon dioxide by diffusion through the body surface (Brusca et al., 2016). The outermost layer of the organism is the epidermis with cilia used for locomotion. The only sensory organ they have is a statocyst, containing a statolith, connected to the 'brain' that helps the organism for balance and orientation.

The genome

The number of genes varies widely between species from 18,000 to 32,000 (Tomiczek, 2017), that is comparable to the number of genes in the human genome.

Xenacoelomorpha contain all 11 animal homeodomain transcription factor classes, suggesting that they were also present in the last common ancestor of bilaterians. This is in contrast with the loss of some classes in higher organisms like *Caenorhabditis elegans* (Hench et al., 2015) and *Drosophila* (Bürglin and Affolter (2016). *Xenacoelomorpha* are the first among the extant metazoans to display medio-lateral axis and bilateral symmetry with the right and the left sides of the body appearing as mirror images of each other. It is worthwhile mentioning that an inverse correlation exists between the complexity of the nervous system and the diversity of 'neurogenic' genes in *Xenacoelomorpha* (Martinez et al., 2017).

Xenoturbella species inherited many ancestral bilaterian peptidergic systems and several clade-specific neuropeptides and preproneuropeptides, but surprisingly most of them are not identified in acoels (Thiel et al., 2018).

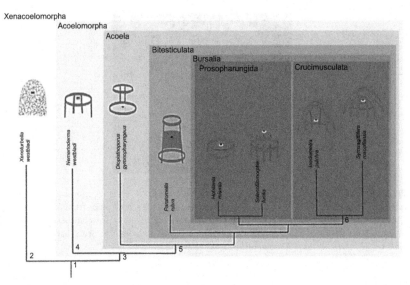

FIGURE 4.20 Schematic organization of the nervous system in different members of the *Xenacoelomorpha*. (1) Putative Xenacoelomorpha ancestor. (2) *Xenoturbella* nerve net located basiepidermally. Presence of a statocyst. (3) Presence of a ring commissure plus a low number of neurite bundles, extending along the anterior–posterior body axis. (4) Presence of a statocyst with two statoliths, typical of the Nemertodermatida. (5) Progressive loss of the basiepidermal location of the nervous system and tendency to possess a second, anterior, ring commissure. The statocyst has one statolith. (6) Replacement of the anterior ring-commissure by bilobed brain, with a dense neuropil, plus one to three anterior commissures. Nervous systems information from: *Xenoturbella westbladi* (Raikova et al., 2000; Israelsson, 2007); *N. westbladi* (Raikova et al., 2004); *D. gymnopharyngeus* (Smith and Tyler, 1985); *P. rubra* (Crezée, 1978); *H. miamia*; *S. funilis* (Crezée, 1975); *Isodiametra pulchra* (Achatz and Martínez (2012)); *S. roscoffensis* (Bery et al., 2010; Bery and Martínez, 2011; Semmler et al., 2010) (current study). *Adapted from a diagram in Achatz, J., Martínez, P., 2012. The nervous system of Isodiametra pulchra (Acoela) with a discussion on the neuroanatomy of the Xenacoelomorpha and its evolutionary implications. Front. Zool. 9, 27, from Perea-Atienza, E., Gavilán B., Chiodin, M., Abril, J.F., Hoff K.J., Poustka, A.J., Martinez, P., 2015. The nervous system of Xenacoelomorpha: a genomic perspective. J. Exp. Biol. 218, 618–628.*

The nervous system

The organization of neurons in the form of a nerve net suggests with high likelihood, that Xenoturbellida represents the basal condition for *Xenacoelomorpha* (Gavilán et al., 2016). Even under the most inclusive definition of the brain or brain-like structure: "the anterior centralization that contains a central neuropile surrounded by a cortex of nerve cells" (Gavilán et al., 2016), Xenoturbellida would represent brainless eumetazoans. Neither the brain nor brain-like structures as nerve rings, ganglia, or neurite bundles are present (Israelsson, 2007; Nakano, 2015), although some earlier studies had suggested that Acoela species *Symsagittifera roscoffensis* possesses a brain structure (Bery et al., 2010) (Fig. 4.20).

(A)   **(B)**

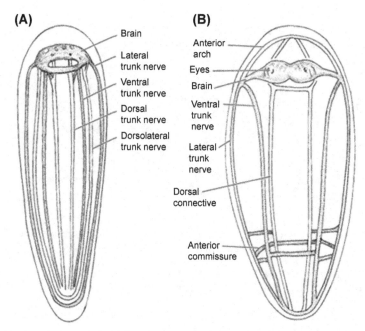

FIGURE 4.21 Comparison of the central nervous systems of (A) an acoel (*Actinoposthia beklemishevi*) and (B) a free-living flatworm (*Gievzstoria expedita*). *From Brusca, R.C., Moore, W., Shuster, S.M., 2016. In Invertebrates. Chapter 9: Introduction to the Bilateria and the Phylum Xenacoelomorpha — Triploblasty and Bilateral Symmetry Provide New Avenues for Anomal Radiation, third ed. Sinauer Associates, Inc., Sunderland, MA, U.S.A. pp. 354.*

Acoelomorpha evolved bilateral symmetry despite the fact that many of them are only in possession of weakly condensed basiepidermal nerve net, similar to the one observed in cnidarians. Their anterior—posterior patterning is determined by expression of BMP and BMP inhibitors, but they develop no dorso-ventral patterning (Martín-Durán et al., 2018) as other bilaterians do. Prevailing opinion is the Xenoturbellida are extant simple organisms resembling the common bilaterian ancestor (Nakano, 2015).

The free living exerted a selective pressure for evolving higher concentrations of neurons and evolution of better organized brains in Acoela, as opposed to the less challenging parasitic life forms (Fig. 4.21).

Notice the difference in the degree of the centralization of the nervous system and formation of the bilobed brain and anterior commissure in the free-living flatworm.

Phylogeny and evolution

Xenoturbellida (both *X. bocki* and *X. westbladi*) is a sister group of all *Acoelomorpha*, but their position in the phylogenetic tree is not firmly established (Nakano, 2015).

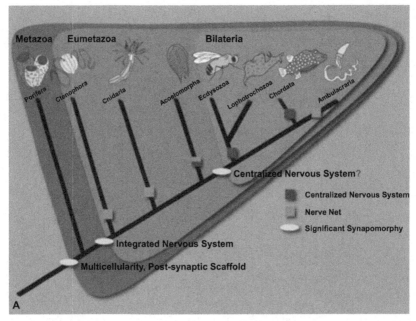

FIGURE 4.22 Evolution of neural characters in the metazoa (A) Protostomes (Ecdysozoan and Lophotrochozoan) and Deuterostomes (Chordate and Ambulacrarian) comprise lineages in which at least some taxa posses centralized nervous systems. Basal metazoan taxa display varying degrees of neural organization ranging from a lack of neural elements (Porifera) to nerve nets (Ctenophora and Cnidaria) to anterior ganglia and multiple nerve cords (*Acoelomorpha*). *From Marlow, H.Q., Srivastava, M., Matus, D.Q., Rokhsar, D., Martindale, M.Q., 2009. Anatomy and development of the nervous system of Nematostella vectensis, an anthozoan Cnidarian. Dev. Neurobiol. 69, 235–254*

Two opposing hypotheses are presented on the origin of *Acoelomorpha*. One, the "acoeloid–planuloid hypothesis," posits that the group is a descendant of an Urbilaterian, whereas the second hypothesis considers the simplicity of their organization, the presence of a basiepidermal nervous system, lack of through gut and anus and lack excretory organs, as a result of the secondary loss (Achatz et al., 2013).

Climbing the tree of evolution, the nervous system generally evolves to ever-increasing forms of organizations from the neural net, to anteriorly concentration of neurons, formation of incipient 'brains", bi-lobed brains and increasing number of neurons in the brain (Fig. 4.22).

Planarians

Planarians are a group of several hundred known species belonging to the subphylum Turbellaria. Of all the known extant bilaterians, they are the first to exhibit cephalization, with concentration in the anterior part of the body

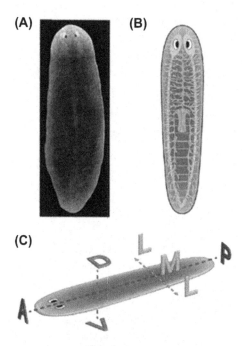

FIGURE 4.23 Planarian anatomy and body axes. (A) Dorsal side of the planarian *Schmidtea mediterranea*. (B) Planarian diagram showing the brain lobes, nerve cords, and secondary nerves (*green [white in print version]*); the two eyes (*black and white*); the gastrovascular tract (*gray*); and the pharynx (*light brown[gray in print version]*). (C) The three main axes of the planarian anatomy: anterior—posterior (AP), dorsal—ventral (DV), and medial—lateral (ML). *From Lobo, D., Beane, W.S., Levin, M., 2012. Modeling planarian regeneration: a primer for reverse-engineering the worm. PLoS Comput. Biol. 8(4), e1002481.*

of neurons in the form of a bilobed brain and two simple eyes - photosensitive ganglia. They also are among the first of extant metazoans to have evolved bilateral symmetry and full-fledged triploblasty, but they did not evolve a body cavity, hence they, along xenoturbellarians, are known as acoelomates. Bilaterian body plan has been described as a triaxial Cartesian system of molecular coordinates (Rentzsch and Holstein, 2018) (Figs. 4.23 and 4.24).

From all the lower known invertebrates, flatworms are the first to have evolved a head region and a brain consisting of two ganglia containing several thousand neurons (Brown and Pearson, 2017). Their central nervous system consists of the brain and a pair of ventral nerve cords (VNCs) (Agata et al., 1998).

Planarians are also the first organisms to have evolved a secretory protonephridial system for regulating the osmotic pressure (removing excess water) and eliminating toxic waste to the outside.

Pigment cups in eyes

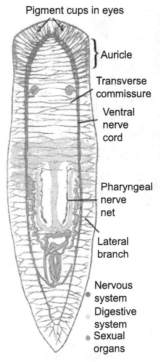

FIGURE 4.24   Overview of the anatomy of the planarian nervous system. Illustration of the nervous system in *Schmidtea polychroa*. (Adapted from Ref. 185) Neural tissues are colored blue, reproductive organs in orange, and the digestive system in yellow. *From Ross, K.G., Currie, K.W., Pearson, B.J., Zayas, R.M., 2017. Nervous system development and regeneration in freshwater planarians. WIREs Dev. Biol. 6, e266.*

Fossil record

Being soft-bodied animals, planarians, just like other platyhelminthes, left no Cambrian fossils. However, recently biologists have discovered what may be the earliest bilaterian nontrace fossil of ichnospecies *Plexus ricei*, a serially divided tubular organism, 5—80 cm long and 5—20 mm wide. It appeared first during the Ediacaran eon, in southern Australia, about 575 Ma, and disappeared in the lower Cambrian ~540 Ma. It has not been possible to assign *P. ricei* to any extant crown group because of the incomplete preservation (Joel et al., 2014). But the close resemblance of modern planarians to some of these Cambrian fossilized imprints suggests that they may have changed little ever since and the study of their morphology and physiology may provide valuable hints on the structure and function of their ancestral state.

The genome

The genome of the planarian *Schmidtea mediterranea* consists of ~20,000 annotated planarian genes of which ~80% have orthologs in humans (Friedländer et al., 2009). The number of predicted transcription factors in *S. mediterranea* (n = 843) is slightly higher than in other Lophotrochozoans (n = 672), or nematodes (n = 725), and is half the number of vertebrates (n = 1866) or mammals (n = 1786) (Swapna et al., 2018). Planarians made a great evolutionary leap, compared to cnidarians, with a comparable number of genes.

The nervous system

Planarians are the lowest of animals with a widely accepted primitive two-lobed brain consisting of 20–30,000 neurons (Inoue et al., 2015) (Fig. 4.25).

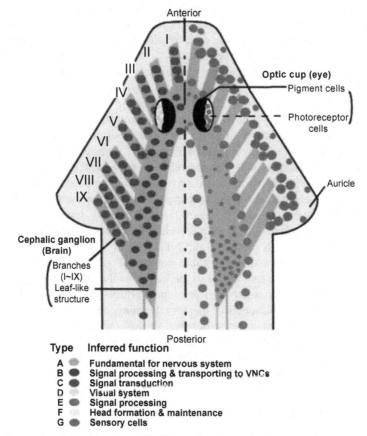

FIGURE 4.25   Cytoarchitecture map of the planarian brain. The represented patterns of each type (A to G) are schematically drawn using different colors at a half side. *From Nakazawa, M., Cebrià F., Mineta, K., Ikeo, K., Agata, K., Gojobori, T., 2003. Search for the evolutionary origin of a brain: planarian brain characterized by microarray. Mol. Biol. Evol. 20, 784–791.*

The centralization of the nervous system was also associated with emergence of eyes as extensions of the brain, specialized visual organs composed of pigment cells that form the optic cup, and opsin-containing photoreceptor neurons.

Planarians are thought to be the first among metazoans to produce brain waves, in the form of electrophysiological oscillations, similarly to higher animals (Aoki et al., 2009). Their brain integrates the photoreceptor, tactile, and chemoreceptor input to generate adaptive systemic and local responses and reflexes (Sarnat and Netsky, 1985).

It has been known for long time that neuroendocrine signals from the CNS control the sexual reproduction in free-living planarians, but a general picture of their neuroendocrine repertoire is only recently emerging. In *S. mediterranea*, 51 genes coding more than 200 peptides are identified. About 85% (44 of 51) prohormone (neuropeptide) genes are expressed in the CNS (brain and ventral nerve cords) of asexual palanarians and the expression of several neurohormones is different in sexual and asexual planarians. Of special importance in the development and maintenance of genital organs in planarians is neuropeptide NPY-8. Extirpation of the brain and nerve cords results in atrophy of testes in these organisms (Collins et al., 2015). The mediator of the action of the neurohormone for the development of the germ cells into mature gametes is its receptor NPYR, but very low ($\sim$50-fold lower) expression of the npyr-1 prevents activation of the NPYR-1 receptor (Saberi et al., 2016).

What determines a planarian's higher position than a cnidarian in the tree of life is neither the number of genes nor the number of neurons: a cnidarian has about 20,000 protein-coding genes (Dupre and Yuste, 2017), 20,000 *Hydra magnipapillata*, and 18,000 *Nematostella vectensis* (Steele et al., 2011). Moreover, about 30% of planarian nervous system—related genes had homologous sequences in the plant *Arabidopsis thaliana* and yeast, which have no nervous system (Mineta et al., 2003). This clearly suggests that these nervous system—related genes were recruited from existing genes previously used for other functions (Steele et al., 2011).

Being in possession of comparable number of genes and neurons with cnidarians, planarians stand higher in the tree of evolution by having evolved triploblasty, bilateral symmetry and developing organs (eyes, auricle, pharynx, digestive system, genital organs), memory, learning, and context-specific behaviors. It is significant to point out that all these novelties coincided with evolution of the bilobed brain, bilateral concentration of 18 neural ganglia in the anterior part of their body, providing them the necessary computational potential for developing structure and functions of greater complexity.

Neural control of behavior

The brain evolution may have been accompanied by the evolutionary emergence of the neural stem cell system, which enabled a drastic increase of the variety of neural cell types that can be produced from a single precursor via asymmetric cell division (Agata et al., 2006).

Planarians are the lowest of animals that evolved a full fledged two-lobed brain consisting of 20—30,000 neurons (Inoue et al., 2015). The centralization was also associated with emergence of eyes as extension of the brain, specialized visual organs, composed of pigment cells that form the optic cup, and opsin-containing photoreceptor neurons, which transform light stimuli into electrical signals that are integrated and processed in the brain.

Neuropeptides are probably the most ancient signaling molecules in metazoans. In platyhelminthes, they appear in the form of proneuropeptides (pNPs). The catalog of *Platynereis dumerilii* neuropeptides is especially rich with 98 pNPs (Conzelmann et al., 2013). The complexity of prohormone expression within the primitive planarian CNS is surprising.

Planarian brain performs integration of multiple external signals to determine adaptive behaviors to the environment (Inoue et al., 2015). Among the various behaviors performed by animals of this group are locomotion, food-finding, feeding, predation, phototaxis, light-avoidance behavior, thermotaxis, positive thigmotaxis (movement toward the food), negative thigmotaxis (movement away of an object), and chemotaxis. Light avoidance behavior is determined, at least partly, by secretion of products of the *1020HH* and eye53 by neurons in a dorsomedial region of the brain during experimental regeneration of the head in planarians (Inoue et al., 2004). Headless planarians and those treated with synaptotagmin in the brain lose chemotactic and thigmotactic behaviors (Inoue et al., 2015).

Planarian brain integrates and processes the sensory input from exteroceptors in specific neural circuits, where behavioral decisions are made (Inoue et al., 2015). The output of the processing determines worm's behavior to lights of various wavelengths, including photophobic responses of the worm to lights of particular wavelengths (Paskin et al., 2014). The planarian food-finding behavior is a chemotactic behavior based on the perception in the food of a few peptides consisting of 10—50 amino acid residues (Miyamoto and Shimozawa, 1985) (Fig. 4.26). To maintain normal brain functions, the planarian brain must have a minimum of 10,000 neurons, so that when planaria lose body mass, as a result of starvation, the number of the neurons in their brain falls under that minimum and they fail to perform their functions including behaviors (Takeda et al., 2009).

The feeding behavior comprises the movement of the animal toward the food, pharynx extension, and food ingestion (Fig. 4.27). The feeding behavior (pharynx extension) is also determined by the brain as demonstrated in experiments of head amputation that prevent planarians from feeding (pharynx extension) behavior and indications that the feeding behavior is regulated by several neuropeptides released by specific planarian brain neurons (Shimoyama et al., 2016). However, animals with RNAi of the prohormone convertase 2 (PC2), a neuropeptide processing enzyme gene, did not perform the pharynx extension behavior, suggesting the possible involvement of neuropeptide(s) in the regulation of pharynx extension.

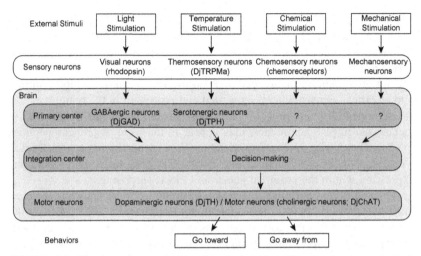

FIGURE 4.26 Neural networks controlling planarian behaviors. Planarians receive many kinds of signals from the outside, through independent types of sensory neurons such as rhodopsin-expressing visual neurons, thermosensory neurons, and chemoreceptor-expressing chemosensory neurons, and these signals are processed by neurons in the brain, such as GABAergic neurons and serotonergic neurons. Thereafter, the various signals are integrated via certain neural networks in the brain to decide a planarian's behavioral strategy. Subsequently, the planarian behaves suitably in response to its conditions by controlling its motor neurons. *From Inoue, T., Hoshino, H., Yamashita, T., Shimoyama, S., Agata, K., 2015. Planarian shows decision-making behavior in response to multiple stimuli by integrative brain function. Zool. Lett. 1, 7.*

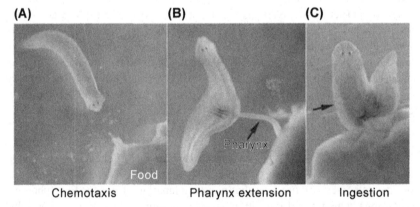

FIGURE 4.27 Planarian feeding behavior consists of three steps. (A) *Chemotaxis*: Planarians move to food (chicken liver). (B) *Pharynx extension*: Planarians extend their pharynx from the ventral side of the body after arriving at food. Pharynx is indicated by a black arrow. (C) *Ingestion*: Planarians ingest food via the pharynx. The color of the intestinal duct changes to red (the color of the food, *black arrow*). *From Shimoyama, S., Inoue, T., Kashima, M., Agata, K., 2016. Multiple neuropeptide-coding genes involved in planarian pharynx extension. Zool. Sci. 33, 311–319.*

(A)

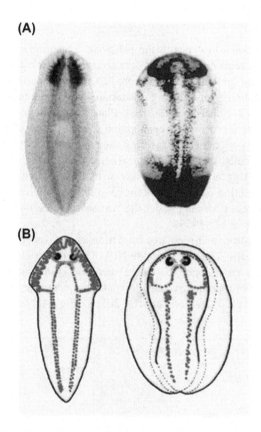

(B)

FIGURE 4.28 Comparison of planarian CNS with the primary CNS of a *Xenopus laevis* embryo. (A) Left panel shows a ventral view of planarian CNS stained by the PH04 gene as a probe. The right panel shows a dorsal view of the primary CNS of a *X. laevis* embryo at stage 17 stained by X-Delta-1 as a probe (reproduced from Chitnis et al. (1995); with permission from C. Kintner). (B) Schematic drawings of planarian CNS and the primary CNS of a *Xenopus* embryo at stage 14. PH04-positive neurons in the planarian and primary neurons of the *Xenopus* embryo are indicated in red. Two eyes and projection routs of visual axons of a planarian, and presumptive eyes and chiasma region of the *Xenopus* embryo are added onto the drawings. *From Agata, K., Soejima, Y., Kato, K., Kobayashi, C., Umesono, Y., Watanabe, K., 1998. Structure of the planarian central nervous system (CNS) revealed by neuronal cell markers. Zool. Sci. 15, 433–440.*

Planarians can also learn as demonstrated in experiments with one of the simplest forms of learning such as the classical conditioning and habituation (Nicolas et al., 2008). Planarians represent the first group of animals displaying associative learning (Mueller, 2002).

An interesting example of the conservation of the stages of evolution of the nervous system during development is the fact that a planarian-like CNS is observed at the stage 14 of the African clawed frog *Xenopus laevis* (Agata et al., 1998) (Fig. 4.28).

Reproduction

There are two known biotypes in the planarian *S. mediterranea*, hermaphroditic sexuals that reproduce by cross-fertilization and asexuals that reproduce by fission, but the general mode of reproduction in planaria is sexual reproduction and sexually reproducing planarians are hermaphroditic (Martín-Durán et al., 2012; Saberi et al., 2016). Planarians lack genital organs. Fissioning represents a process of regeneration (Neuhof et al., 2016). Given the fission is generally asymmetric, it is suggested that epigenetic changes in cells, including nerve cells, convey to the regenerated individuals different epigenetic traits, affecting not only the behavior of these individuals but also the memory and evolution of planarians. In conditions of higher population densities, they suppress fissioning under the "probably neurohormonal" influence of the brain (Best et al., 1969).

The brain controls and regulates formation of the somatic germ lines and gonads in planarians. The neuropeptide NPY-8 is a key prohormone produced by the planarian central and peripheral nervous system that performs the systemic control of reproductive development in a neurohormonal fashion via its receptor, NPYR-1 (Saberi et al., 2016) (Fig. 4.29), by stimulating

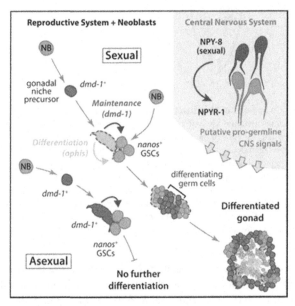

FIGURE 4.29 Schematic of the developmental mechanisms involved in planarian testis formation. dmd-1[+] cells in both sexual and asexual worms are required for specification of nanos[+] GSCs (germline stem cells). In sexual planarians, these dmd-1[+] cells express *ophis*, which is required for further differentiation of GSCs into mature gametes. NPY-8 signaling, which occurs in the CNS, systemically promotes later stages of germ cell maturation. *NB*, neuroblasts. *From Saberi, A., Jamal, A., Beets, I., Schoofs, L., Newmark, P.A., 2016. GPCRs direct germline development and somatic gonad function in planarians. PLoS Biol. 14(5), e1002457.*

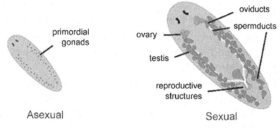

FIGURE 4.30 Schematic of the reproductive system in the two biotypes of *Schmidtea mediterranea*. Sexual planarians (right) develop a complete reproductive system, including mature gonads and accessory reproductive organs. The asexuals (left) contain only presumptive gonads with PGCs (primordial germ cells). *From Saberi, A., Jamal, A., Beets, I., Schoofs, L., Newmark, P.A., 2016. GPCRs direct germline development and somatic gonad function in planarians. PLoS Biol. 14(5), e1002457. From Oviedo, N.J., Morokuma J., Walentek, P., Kema, I.P., Gu, M.B., Ahn, J-M. et al., 2010. Long-range neural and gap junction protein-mediated cues control polarity during planarian regeneration. Dev. Biol. 339, 188−199.*

differentiation of the germ cells. During the asexual reproduction, suppression of the neurohormone secretion occurs (Collins et al., 2015).

Secretion of the protein CPEB-2 in the brain of the model planarian *S. mediterranea* is necessary for spermatogenesis and the development of ovaries and accessory reproductive organs. Experimental evidence suggests that by promoting translation of the brain neuropeptide Y-8 (NPY-8), CPEB-2 promotes sexual maturation and formation of reproductive organs in planaria (Rouhana et al., 2017) (Fig. 4.30).

Morphogenesis and organogenesis

Planarians can change the number of cells in the body proportionally to accidental/experimental changes in length and mass of the body via apoptosis (programmed cell death), proliferation, and differentiation of neoblasts, thus "maintaining a constant ratio of different cell types" (Takeda et al., 2009). Regeneration experiments demonstrate they are in possession of "some form of counting mechanism" that has the ability to regulate both the absolute and relative numbers of different cell types in complex organs such as the brain during cell turnover, starvation, and regeneration" (Takeda et al., 2009) and reestablish it in the process of regeneration. Even when cut to small pieces, planaria forms miniature individuals that maintain the proportions of the body, organs, and ratios between the different types of cells.

To the question "Do any organs have the capacity to determine the size of other organs?", based in their own experiments investigators respond affirmatively that "the brain may have the capacity to determine the size of other organs in planarians" (Takeda et al., 2009). If such a mechanism evolved in planarians, there is no visible reason that evolution would later wipe it out of the rest of higher bilaterians, and in extant bilaterian clades. Indeed, adequate experimental evidence shows that the brain determines the body mass in invertebrates (Rulifson et al., 2002; Colombani et al., 2005; Truman, 2006; Krieger et al., 2004) and vertebrates (Adams et al., 2001; Levin and Dunn-Meynell, 2000; Baeckberg et al., 2003).

Experiments of regeneration of the anterior part of the planarian body also show that the nervous system via neurally derived signals regulates animal's morphogenesis, differentiation of pharyngeal, and cephalic structures. It is concluded that this together with the results of the previous experiments of regeneration in annelids and amphibians "suggests a conserved role of the nervous system in pattern formation during blastema-based regeneration" (Cebrià and Newmark, 2007).

Planarian regeneration and development

The process of regeneration is relevant to the planarian development because it essentially implies reactivation of developmental processes to restore the missing parts. The regeneration of the amputated parts of planaria indicates that the parts have a sense of the loss of structures and remember the structure of the missing part. The parallelism between the regeneration and development has been repeatedly emphasized earlier (Solana, 2013; Neuhof et al., 2016).

Planarians can regenerate after virtually all amputation scenarios. Even a piece as small as 1/279 of the adult planarian can regenerate (Newmark and Alvarado, 2002). This requires a robust system that instructs stem cells to correctly replace missing tissues. Head regeneration starts with head-versus-tail decisions at amputation sites, which involves temporarily resetting positional information.

Experimental evidence that a subpopulation of neoblasts (planarian multipotent stem cells), the v-neoblasts, or neoblasts of the neural lineage reside adjacent to the worm's brain suggests that they may be *bona fide* neural stem cells (Brown and Pearson, 2017). These neural neoblasts are vital for the adult neurogenesis in planaria (Brown and Pearson, 2017), especially in the process of regeneration, when these neuron precursors participate in *de novo* formation of the brain from a tiny part of their body (Cebrià, 2016). Neoblasts comprise ~25% of cells in the adult animal (Friedländer et al., 2009). It is suggested that brain evolution was accompanied by emergence of a neural stem cell system via asymmetric cell division, which facilitated differentiation of various neuron types in planarians (Umesono and Agata, 2009).

In the process of regeneration, neoblasts begin proliferating and migrate to the amputation sites to form the regeneration blastema (Reddien and Sánchez Alvarado, 2004). At the site of injury forms a blastema, regeneration bud of neuroblasts, which receives from the CNS instructions on what part of the body is missing and the morphology it must develop (anterior or posterior part of the body). The instructions are provided to the blastema in form of neural signals from the ventral nerve cord (VNC) or via the gap junctions. The neural signals contain information on the existence of the head very early, ~6 h after the amputation, after which the regenerating fragment begins developing the posterior part of the body. When no signals from the CNS reach the blastema, by default the regenerating fragment starts developing the anterior part of the body, including the brain. The instructive role of the nervous system is corroborated by the fact that experimental prevention of CNS signals to reach blastema by blocking the gap junction communication leads to formation of an ectopic head (Oviedo et al., 2010).

Each of the amputated parts of the planarian body must determine which and how much of tissue have to be produced: the part of the missing body (anterior or posterior) and the right arrangement of cells of different types in regenerating parts (Owlarn and Bartscherer, 2016).

The instructive role of the nervous system in planarian regeneration

Planarian regeneration begins with the proliferation planarian stem cells or, as commonly called, neoblasts. Their proliferation is induced by neuropeptides P and K, which, via their specific membrane receptors, show mitogenic activity even in nanomolar quantities (Baguñà et al., 1989). It is observed that by modulating gap junction between cells and blocking neural signals, ectopic blastemas are induced during regeneration, indicating the existence of the long-range instructive regulation of the nervous system in communication between neuroblasts and the regenerating fragments (Oviedo et al., 2010).

Blastemas on two sides of amputation are instructed which part of the body to build (the head or tail side) by neural signals coming via uninterrupted VNCs and innexin proteins via functional GJC (gap junction communication). When blastemas are treated with octanol, a disrupter of VNC integrity and GJC inhibitor alterations in anterior—posterior polarity occur (Oviedo et al., 2010; Emmons-Bell et al., 2015) (Fig. 4.31). The instructive role of the nervous system in regeneration is well-known in other clades and even in vertebrates (Kumar and Brocke, 2012).

The experimental observation that reduction of brain ganglia impairs blastema formation also suggests a causal relationship between the brain and the regeneration in planaria (Fraguas et al., 2014) (Fig. 4.32).

Neural correlates and driving forces of the Cambrian explosion

The last two decades saw the demise of the long-held notion on existence of a correlation between the increase of the number of genes and complexity of the genome with the phenotypic complexity of living metazoans (Schmidt-Rhaea, 2007). This favored the idea that not the number of genes or evolution of new

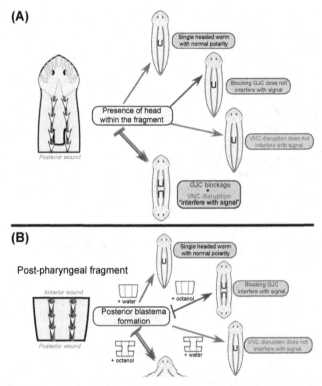

FIGURE 4.31 Schematic model of gap junction–mediated and neural signals during regeneration. (A) Postpharyngeal amputation generates a regenerating worm with preexisting head and a posterior wound. Our data suggest that long-range signals — informing about the presence of a head within the fragment - travel from anterior (brain in *yellow [white in print version]*) to the posterior wound (*white arrowheads*) through two pathways: ventral nerve (VNC, *red line [gray in print version]*) and GJ-mediated signals associated with VNC (*blue line [black in print version]*) and parenchymatic cells (*light blue [dark gray in print version] background*). In all, untreated animals or animals exposed to GJ blocker (octanol) and animals with VNC disruption regeneration of the missing posterior area is recreated without polarity problems. However, in animals where both octanol treatment and VNC disruption take place, signals coming from anterior areas are disrupted, leading to abnormalities in polarity (bipolar animals). (B) Blastema fate determination in regenerating post-pharyngeal fragments with anterior and posterior-facing wounds. Instructive signals from anterior wound travel to posterior wound through the same pathways as in "A" but in this case the information carried is to inform about the anterior blastema formation that instruct neoblasts to form a posterior blastema. In animals exposed to octanol this information is altered and animals regenerate bipolar heads. In the case of VNC disruption alone most animals regenerate without problems. If both VNC are disrupted and GJ are inhibited, animals regenerate four heads. Specific cuts for VNC disruption are represented for each case. *From Fraguas, S., Barberan, S., Iglesias, M., Rodriguez-Esteban, G. and Cebria, F., 2014. egr-4, a target of EGFR signaling, is required for the formation of the brain primordia and head regeneration in planarians. Development, 141, 1835–1847.*

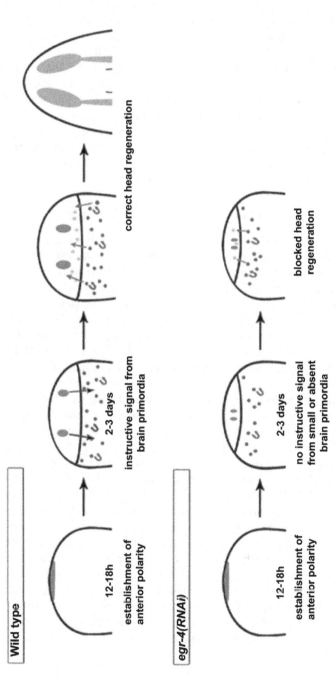

FIGURE 4.32 Proposed model illustrating the requirement of brain primordia for head regeneration in planarians. In wild type the brain primordia would send some signal to the stump to promote the proliferation, migration or differentiation of the neoblasts to allow blastema growth and head regeneration. In the absence of a proper brain primordia after egr-4 RNAi, the lack of such putative inducing signal would explain the inhibition of head regeneration. *Blue (gray in print version)*, anterior pole; *red dots (black in print version)*, proliferating neoblasts; orange dots, differentiating neoblasts; purple dots, brain primordia; green, mature CNS. *From Fraguas, S., Barberan, S., Iglesias, M., Rodriguez-Esteban, G. and Cebria, F., 2014. egr-4, a target of EGFR signaling, is required for the formation of the brain primordia and head regeneration in planarians. Development, 141, 1835–1847.*

genes but changes in the patterns of their expression are the source and the driving force in the evolution of living world. This idea was undermining the fundamental neo-Darwinian tenet on mutations or changes in the genetic information as the main source of evolutionary change. Biologists, thus, were faced with a serious theoretical problem: if changes in genes or in the genetic information are not involved in the evolution of living forms, then the mechanism and the source of information for evolutionary diversification are anything but a genetic mechanism. This general explanatory conundrum and the weak relevance of the ecological and environmental factors as possible triggers of the Cambrian explosion (see Section Triggers of the Cambrian explosion: hypotheses in Chapter 2) logically stimulated search for other causes of this enigmatic and unique event in the history of life.

Selective pressure for centralization of the nervous system

The fossil record indicates that the Ediacaran eon was generally a predationless world, but by the end of the Ediacaran, with the appearance of the first bilaterian predators, both predators and their preys had to evolve behavioral, morphological, and physiological features that would help them to survive in an ecological environment that was becoming more and more complex.

Monk and Paulin believe that the onset of predation by the end of the Ediacaran and early Cambrian arose a selective pressure to evolve complex behaviors and more elaborated nerve systems (Monk and Paulin, 2014). In the new increasingly unpredictable environment, animals need to gain and process the greatest possible amount of information about their biotic environment to survive and better adapt to the environment. The most sparing way to this was by increasing the processing and integrating power of the nervous system, by evolving its centralization, i.e., concentration of neurons in the anterior part of the body into brain-like or brain structures in a process that minimizes the energy by reducing the cost of connections through shortening the travel distances of signals or the total neuronal connection length, the so-called "saving wire" rule (Chklovskii, 2004; Gavilán et al., 2016). Centralization of the nervous system was also accompanied by a concentration of sensory organs in the vicinity of the CNS in a mutually supporting process of concentration in the head region. The increase in the number of neurons and the accelerated centralization of the nervous system, by enabling formation of energetically less expensive synaptic connections and by increasing their number of connections, exponentially increased computational power of the nervous system and the capability of animals to process and integrate external and internal stimuli. Centralization of the nervous system enabled metazoans to thoroughly monitor not only the external environment but also the state of the living system.

Centralized nervous system in cambrian fossils

The oldest known fossil with a centralized nervous system is that of a crustacean-like organism, *C. kunmingensis*, found in South China (Fig. 4.33).

(A)   **(B)**   **(C)**

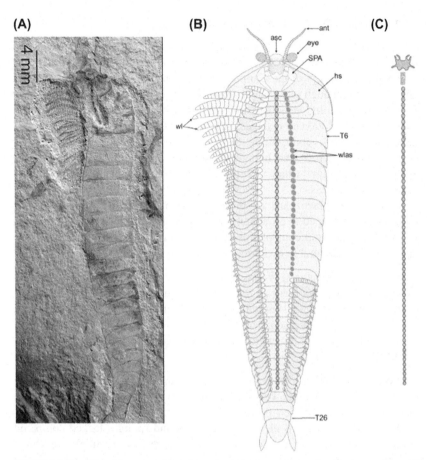

FIGURE 4.33   *Chengjiangocaris kunmingensis.* (A) Photo of the fossilized organism. (B) Morphological reconstruction of *C. kunmingensis.* Complete exoskeletal morphology in ventral view showing arrangement of walking legs (wl) and their attachment sites to the body (wlas) relative to the preserved VNC *(purple [light gray in print version])* and tergites (T*n*); note that only T1–T5 have a one-to-one correspondence with the walking legs. (C) Overall view of the CNS, including the VNC and dorsal brain; the gap between the VNC (ventral nerve cord) and brain reflects lack of paleontological data pertaining this region. *SPA*, specialized postantennal appendage; hs, head shield; ant, antennae. *From Yang, J., Ortega-Hernández, J., Butterfield, N.J., Liu, Y., Boyan, G.S., Hou, J-b. et al., 2016. Fuxianhuiid ventral nerve cord and early nervous system evolution in Panarthropoda. Proc. Natl. Acad. Sci. U.S.A. 113, 2988–2993.*

It has about 80 legs and the body length varies from several to 15 cm. Preserved in the fossil is the postcephalic neurological structure and the ventral nerve cord, with interconnected metameric ganglia, whose size is reduced toward the posterior end of the body. The tripartite brain with olfactory and optic lobes is not preserved (Yang et al., 2016).

Centralization of the nervous system

Most hypotheses of the Cambrian explosion look at external environmental agents as possible triggers of the explosive radiation of bilaterians at that time (see Section Triggers of the Cambrian explosion: hypotheses, Chapter 2). As already said, although those agents must have played a role in selection of the Cambrian fauna, obviously they are not capable of producing or increasing the frequency of evolutionary changes, i.e., changes that would be inherited across generations. While contributing to the knowledge of the selective role of environmental factors, the hypotheses do not deal with the causes of the evolutionary change, which are prerequisite for, and the lifeblood of the evolutionary process.

Sponges and placozoans are in possession of the basic metazoan genetic toolkit but are trapped in an evolutionary dead end for more than 600 million years. Cnidarians and ctenophores when compared with sponges and cnidarians show a considerable evolutionary progress mainly in

- the appearance tissues, groups of cells of a certain morphology and function, which defines them as tissue-grade organisms,
- radial symmetry with a central axis as the main determinant of morphology, which sponges and placozoans lack, and
- fixed adult body plans, which lack in sponges and placozoans (Finnerty, 2001).

No relevant differences in the size and composition of their genome, the genetic toolkit, or the number of cell types have been advanced as possible explanatory factors of their position in the tree of evolution.

The most conspicuous element placozoans and sponges lacked compared to their closest relatives, cnidarians and ctenophores, is a particular cell type, the neuron. While this does not imply that the neuron is a causal factor of the evolutionary progress, it is worth remembering that of all the cell types in cnidarians, only neurons are essentially involved in induction of cell differentiation, development, and metamorphosis in this phylum.

But if the neuron and the diffuse nervous system of extant cnidarians determine their development, there is no visible reason to doubt that the neuron and the nervous system were essentially involved in their emergence. Indeed, there is no other way for the evolutionary changes/transitions to occur but via the development, via changes in developmental pathways, which modern biology indicates that are highly conserved in evolution. But cnidarians too are a very conservative clade that relatively less diversified and remained "living fossils," essentially unchanged for at least 600 million years.

Although there are about 9000 living cnidarian species, all of them show very little morphological diversity and have a common Bauplan (Ryan et al., 2006), while the sister group of bilaterians generated several million extant and extinct species, grouped in ~30 phyla with ~30 different basic body plans. Bilateria comprise about 99% of the extinct metazoans (Krämer-Eis et al., 2016), although relatively little changes have taken place in their genome during about 600 million years of evolution, and it is estimated that about 85% of genes of the extant bilaterian clades were present in the Urbilateria, their last common ancestor (Srivastava et al., 2010).

While Cnidaria and Ctenophora share the same genetic toolkit with bilaterians (Putnam et al., 2007), what is that 'privileged' *bilaterians alone* to produce enormously diversified metazoan forms within a span of ~10–20 million years during the Cambrian explosion? Why, with the same basic genetic toolkit, cnidarians and ctenophores got stuck in an 'eternal,' 600–800 million year long, evolutionary stasis?

The fact that cnidarians and bilaterians shared the same environment indicates that question "To evolve or not evolve" was determined inherently/constitutively rather than externally; hence, as mentioned earlier, our quest for causal agents of the Cambrian explosion has to focus on inherent properties of organisms of the two groups.

A parsimonious way to get some clues on the possible intrinsic causes of the Cambrian explosion would be to look whether the Cambrian explosion coincided with evolution of any relevant structures in eumetazoans, whether that structure continued to play a role in their later evolution or whether any organ plays the role of inducer in development.

Evolution of bilaterians was associated with an increase in complexity compared to cnidarians. They feature the triploblasty, anterior concentration of the nervous system, bilateral symmetry, and emergence of new cell types (Erwin, 2009). It seems reasonable to look among these novelties and organs for the possible trigger of the Cambrian evolution. Bilateral symmetry is only a concept or spatial characterization of a structure but not a material structure that can store, transmit, or generate information, hence cannot be taken into account as a possible causal factor in bilaterian evolution. The evolution of the new cell types as well cannot be considered, but for a different reason: it is not a typical bilaterian feature because sponges, one of the two most primitive nerveless metazoans, have more cell types than bilaterian Acoela. Similarly, the evolution of the triploblasty is not typical because various cnidarian taxa evolved local triploblasty with a mesodermal layer consisting of striated and smooth muscles (Boero et al., 1998, 2007), and even it is suggested that diploblasty in cnidarians is derived from ancestral condition of triploblasty (Martindale et al., 2004).

Thus, we are left with a single choice, the centralization of the nervous system as a possible source of information for the evolution of ever-increasing complexity, which implies ever-increasing amount of information invested for erecting complex animal structures.

But before dealing with that, let us see whether any correlations exist between the genome and the evolution of complexity in bilaterians.

Evolution of the genome and the complexity of bilaterians

The bilaterian evolution is characterized by increased structural complexity, but no correlation has been possible to find between the structural complexity and the genome size or number of genes in animals. The number of genes in most animals varies between 18,000 and 32,000 (Tomiczek, 2017). So, e.g., while one of the simplest known worms, C. *elegans*, has ~20,000 protein-coding genes (Hodgkin, 2001), one of the most complex organisms, humans, have less than 20,000 protein-coding genes (Ezkurdia et al., 2014), or about 4000,000 genes less than a dog. In another example, mice have ~30,000 genes (Mouse Genome Sequencing Consortium, 2002) and humans ~20,000 (Dolgin, 2017), but even when they share 99% of their genes, their morphologies, physiologies, and behaviors are so different.

An inverse relationship is observed between the number of the bHLH proneural genes and the degree of the complexity of the nervous system: X. *bocki*, with a very simple diffuse intraepidermal neural net, has 33 proneural bHLH (basic helix—loop—helix) genes, i.e., twice the number of bHLH genes of the Acoela species, with bilobed brain, S. *roscoffensis*, a fact that is suggested to be consequence of the loss of these genes in xenoturbellids (Perea-Atienza et al., 2015).

Similarly, the number of the cell types is not related to the degree of the bilaterian structural complexity; a nerveless sponge has at least 16 cell types (without mentioning subtypes) (Leys, 2015), i.e., more than bilaterian planarians, brained animals, with 13 cell types (Wurtzel et al., 2015).

According to the prevailing neo-Darwinian opinion, it would be expected that the transition from the basiepidermal neural net of xenoturbellids to the brain structures observed in Acoela should be characterized by acquisition of new/mutated genes or by an increase in the gene complement of the latter group. Paradoxically, the contrary did occur. Structural complication of the nervous system in Acoela was accompanied by reduction of the number of genes involved in building the bilobed brain connected to two ventral nerve cords, in Acoela (Perea-Atienza et al., 2015).

Strong correlation between the centralization and growth of the nervous system and the evolution of complexity

A clear correlation is observed between the increased complexity in eumetazoans, on the one hand, and

1. The evolution of the centralized bilateral brain, and
2. The increase in the number of neurons in the brain.

A striking positive correlation is observed between *the centralization of the nervous system and the degree of morphological and behavioral complexity*: the centralized nervous system with the same number of neurons is able to perform more complex behaviors and developmental functions than the diffuse neural net because it is able to form more synaptic connections. So, the brainless cnidarian, *Hydra vulgaris* (*H. attenuata*), has 5600 neurons distributed throughout the animal body in the form of a diffuse neural net, whereas a more complex 'brainy' rotifer, *Asplanchna brightwellii*, has only 200 neurons; the bilaterian nematode *C. elegans* is certainly more complex than a cnidarian, but it has just 302 neurons, which form 7500 synaptic connections, whereas the tunicate, *Ciona intestinalis* (sea squirt) larva with only 231 neurons, has a greater number of synaptic connections (8617 in the brain only) (list of animals by number of neurons. Available in Internet https://howlingpixel.com/i-en/List_of_animals_by_number_of_neurons).

If not the number of genes or the number of neurons, what is that makes *A. brightwellii*, *C. elegans*, and *C. intestinalis* incomparably more complex organisms than a cnidarian? The only obvious difference I can see is that in these lower bilaterians the nervous system is centralized to form small brain structure of higher computational power. The idea that this may have been determined by the "unique radial symmetry" of cnidarians (Satterlie, 2018) is not supported and sterile. The only visible difference and the most parsimonious hypothesis that could account for the fact is the *centralization* of the nervous system in bilaterians.

Humans and chimpanzees diverged from their common ancestor only ~6 million years ago and their genetic difference is only 2%–4%, but their anatomic, physiologic, and behavioral characters have changed disproportionately. The change is not paradoxical when we consider that in the meantime the number of neurons was tripled in humans (8.6×10^{10}) compared to chimpanzees (2.8×10^{10}), although they have an almost equal number of genes. It is estimated that during the last 4 million years, an evolutionarily short period of time, the number of neurons in the brain of *Australopithecus* increased from 27 to 35 billion neurons, in *Paranthropus* to 50–60 billion *in Homo rudolfensis* and *Homo antecessor* to 62 billion in *Homo erectus* to 76–90 billion in *Homo heidelbergensis* and *Homo neanderthalensis* (Herculano-Houzel, 2012) to ~100 billion in *Homo sapiens*.

However, even the absolute number of neurons cannot be a fair indicator of the computational potential of the brain because the number of synaptic connections per neuron varies among various animals and even among neurons of the same species. Besides, in the course of evolution, the neuron acquired new structural properties; most neurons in cnidarians and ctenophores do not form dendritic spines. However, some neurons associated with complex sensory inputs such as photoreceptor terminals of motile cnidarians and neurons for rapid prey capture form invaginating synapses. While this type of synapse is conserved in photoreceptor terminals and some mechanoreceptor synapses of higher animals, with the advent of bilobed-brain bilaterians and more complex neural connections, noninvaginating synapses were those that evolved and diversified rapidly (Petralia et al., 2016).

Relevant to the possible role of the CNS in the increased complexity in bilaterians is the fact that it is the first organ system to develop in the bilaterian embryos, although other organ systems (circulatory and excretory) should intuitively be formed first during the embryonic life. As will be later demonstrated, the appearance of the operational CNS at the phylotypic stage as the first organ system to develop is not accidental, but related to its role in organogenesis.

Even a glance on the evolution of organs, organ systems, and other metazoan structures shows that because of this crucial role in development, the neuron and nervous system/brain is the only *organ system* that after evolving is never lost in the course of evolution (Fig. 4.34) (the conservation of

FIGURE 4.34 Select morphological characters. *Dark blue bars (dark gray in print version)* indicate characters that are present, and light blue bars indicate clades with mixed presence/absence of characters. *From Dunn, C.W., Giribet, G., Edgecombe, G.D., Hejnol, A., 2014. Animal phylogeny and its evolutionary implications. Annu. Rev. Ecol. Evol. Syst. 45, 371–395.*

mesoderm, the third embryonic layer, a characteristic feature of bilaterians, is self-explanatory: from it develop muscles, bones, circulation system, blood cells, etc., and the basement membrane in the form of sheets of extracellular matrix is necessary for separating different tissues and organs).

Now, to summarize the above evidence: of all the main features of the transition to bilaterality (emergence of triploblasty, bilateral symmetry, and emergence of new cell types) (Erwin, 2009), the genome and new organs emerging during the Cambrian in bilaterians, only the brain, increase in its size and complexity, shows a clear positive correlation with the evolution of the metazoan behavioral and structural complexity.

Centralization of the nervous system and the exponential increase of computational power

To ask about the trigger or the detonator of the Cambrian explosion is another way of asking about the source of the epigenetic information embodied in the structure, function, and behavior of the Kingdom Animalia that emerged during that time. The thesis developed herein is that the unprecedented and unique burst of diversity of forms during the Cambrian explosion implies the rise of a huge wave of biological information embodied in the structure of the myriad of new species and higher taxa that emerged within that geologically "blink of an eye." This was an information different from the genetic information, *whose only* scientifically known function is the biosynthesis of proteins.

The evolution of the neural net into a CNS, or the brain, may be seen even within the extant clades of the phylum *Xenacoelomorpha* comprising genus *Xenoturbella* and subphylum *Acoelomorpha*. Until recently the genus *Xenoturbella* was considered to consist of only two species, *X. bocki* and *X. westbladi*, but four new species (*Xenoturbella hollandorum, Xenoturbella monstrosa, Xenoturbella churro, and Xenoturbella profunda*) were proposed to belong to the genus in 2016. They possess no brain or nerve rings, but only a simple basal intraepidermal diffuse nerve net, which may be the ancestral condition of the phylum. Species of the order Acoela have evolved a centralized nervous system. So, e.g., in the family of *Solenofilomorphida*, the nervous system is displaced from the epiderm, below the body wall musculature, and forms one to two commissures in the anterior part of the body and eight long neurites. In the class *Crucimusculata,* the nervous system forms a complex ganglionic brain (Gavilán et al., 2016).

However, the centralization in Acoela is rudimentary and their nervous system displays high plasticity among different Acoela species (Gavilán et al., 2016). The concentration of neurons in the anterior part of the body made it possible for acoels to evolve a general bilateral body, but they failed to develop a real bilaterian head with mouth and sense organs. The only sense evolved in many Acoela species are "eyespots," consisting of epidermal pigment cells, but most Acoela evolved no mouth, *bona fide* digestive tract, organs of reproduction, and protonephridia.

The nervous system of Acoela is very small. So, e.g., a simple species of Acoela, *Isodiametra pulchra*, has an estimated 1000 neurons (Achatz and Martínez, 2012). The small number of neurons, the rudimentary centralization, and the resultant low computational power of the nervous system of acoels correlates with the low level of complexity of their structure compared to higher bilaterians. Some studies have concluded that only "some members of *Acoela*, particularly those belonging to the clade *Crucimusculata*, have evolved a condensed bilobed brain" (Perea-Atienza et al., 2015).

The most generally recognized evolutionary "purpose" of the nervous system is to receive information about the environment, to process and respond to it with adaptive behaviors, but the nervous system is also responsible for physiological adaptations based on its role as controller of the function of all other organs and parts of the animal body. The function of control implies as a *conditio sine qua non* the role of the nervous system as a monitor of the state of the system in general. It not only monitors the state of the system at all the levels, down to the lower molecular level, but also receives a continuous flow information on the state of the system by virtue of the pervasive presence of nerve endings throughout the animal body, down to the cell level. Via afferents, the CNS is 'bombarded' with a continuous flow of information on the state of the living system. The input is integrated, processed, and compared with the species-specific set points, which represent physiological norms for thousands of variables. The comparison allows the CNS to determine abnormal changes unavoidably occurring throughout the body and via signal cascades send instructions to target cells/tissues/organs for restituting the damaged structures or abnormal states.

Via afferent pathways, the CNS receives a continuous stream of information. The human brain is capable of processing and storing a volume of information of between 100 trillion to 100 quadrillion bits of information (about 100 billion neurons and each neuron with 1000–10,000 connections whereas a Purkinje neuron forms up to 100,000 connections) per second.

The selective pressure for receiving and processing more and more information on the external and the internal environments led to the exponential increase of the computational power that offered the centralization, the increase in the number of neurons, and formation of 1000–10000 connections per neuron by creating in the human brain quadrillions of pathways through which the afferent and efferent information can travel. The computational power of the higher animals, including humans is not at all excessive or exuberant, if we keep in mind the tremendous complex tasks the nervous system has to carry out for morphogenesis and organogenesis during the development and for maintaining for long periods of time such extremely improbable nonequilibrium structures as higher metazoan organisms (Cabej, 2012b, 2018b).

The process of centralization began with the concentration of neurons in the anterior part of the body followed by formation of neuronal ganglia/brains. The centralization was selected positively during the evolution of animals because of the advantages it offered by shortening the distances for connections between neurons and especially by increasing the computing potential of the nervous system. Brought together in a delimited space, neurons could create more, and energetically less expensive, synaptic connections (1000 to 10,000 connections per neuron in human brains) between dendrites of a neuron with axons of other neurons. With each of thousands of dendrites per neuron serving as computational subunits (Smith et al., 2013), the computational power of the brain consisting of many billions of neurons increased exponentially. Generally, the accelerated multiplication of the computational power of the centralized nervous system was result of the increased synaptic proteome complexity, increase of the number of neurons and the increased anatomical connectivity (Emes et al., 2008; Emes and Grant, 2012).

In humans, even under minimal contact with the external environment and in the absence of problem-solving duties during the sleep the activity of the brain increases, although the conscious activity related to sense perception ceases. This is not paradoxical because the unconscious brain is operative during the sleep for it has to maintain within normal limits numerous physiological and biochemical parameters in body fluids and restitute the unavoidably damaged structures during the day. This seems to be a reasonable explanation why the overall brain activity does not decrease during sleep.

The role of the number of neurons in animal behavioral skills may be illustrated by observations that triploid and tetraploid newts, which have 50% −70% the normal number of neurons, spend two to three times more time to learn a maze than the normal newts do. Administration of growth hormone in immature rats and frogs, by increasing the number of neurons by 20−60%, improves their performance on avoidance conditioning tasks (Williams and Herrup, 1988).

Evolution of the centralization of the nervous system

The emergence of the centralized nervous system is among the most important innovations in the history of life (Gavilán et al., 2016) and a great transition in the evolution of bilaterians. Based on the fossil record, it is suggested that the central nervous systems evolved after the cnidarian-bilaterian split with the advent of predation: "animals evolved spiking neurones soon after they started eating each other" (Monk and Paulin, 2014).

However, the question on whether the centralization occurred only once or more times during their evolution is still unsettled. Comparison of data on the development and spatial organization of the CNS of extant invertebrates and vertebrates suggests that the Urbilateria, the last common ancestor of bilaterians, had a CNS, implying that the nervous system evolved only once, hence

"It is highly unlikely that precisely this mediolateral order and overlap in expression of orthologous genes in the CNS neuroectoderm should evolve twice independently" (Denes et al., 2007).

The similarities in the neurodevelopment and the CNS molecular architecture of extant taxa also suggest that the central nervous system arose only once in the Urbilateria (De Robertis, 2008; Arendt et al., 2008; Strausfeld and Hirth, 2013; Bailly et al., 2013) and gene expression studies in the annelid *P. dumerilii* suggest that "complex brain patterning programs but even individual neural cell types from corresponding progenitor regions are conserved among protostomes and deuterostomes" (Bailly et al., 2013) (Figs. 4.35 and 4.36).

The "single origin" hypothesis is contradicted by the fact that the bilaterian outgroups, ctenophores and cnidaria, have diffuse nerve nets. Besides, the same diffuse organization is also observed in the nervous system of extant deuterostomes and protostomes (acoels, nemertodermatids, xenoturbellids, and chaetognaths) suggests that the ancestral deuterostome also had a diffuse basiepithelial nerve net (Lowe et al., 2003). Moreover, there is evidence suggesting that the protostome-deuterostome ancestor also had a nerve net that was diffuse and basiepithelial, with an evolutionary tendency to become internalized and anteriorly concentrated (Holland, 2003).

The alternative hypothesis posits that centralization of the nervous system occurred many times, at least 5–9 times (in lineages leading to chordates, arthropods, nematodes, annelids, and molluscs) (Northcutt, 2012) (Fig. 4.37).

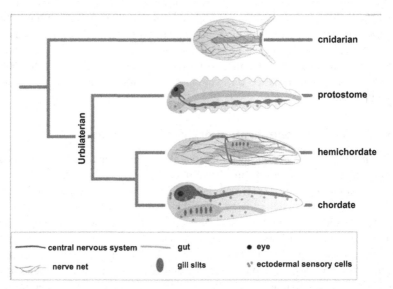

FIGURE 4.35 Comparison of metazoan body plans. A typical cnidarian polyp, a generalized protostome, hemichordate and chordate and their phylogenetic relations are shown. Special attention is given to nervous systems and neural structures of the respective animals. *From Holland, L.Z., Carvalho, J.E., Escriva, H., Laudet, V., Schubert, M., Shimeld, S.M. et al., 2013. Evolution of bilaterian central nervous systems: a single origin? EvoDevo 4, 27.*

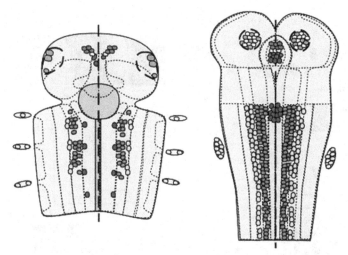

FIGURE 4.36 Conserved neural cell types in annelid (protostome) and vertebrate (deuterostome). The neuron types emerging from homologous regions in the molecular coordinate systems in annelid and vertebrate and expressing orthologous effector genes are marked with the same color. Homologous cell types include the molecular clock cells positive for *bmal* (*dark green [light gray in print version]*), ciliary photoreceptors positive for *c-opsin* and *rx* (*white*), rhabdomeric photoreceptors positive for *r-opsin*, *atonal* and *pax6* (*yellow [white in print version]*), vasotocinergic cells positive for *nk2.1*, *rx* and *otp* (*orange [dark gray in print version]*), serotonergic cells positive for *nk2.1/nk2.2* (*red [black in print version]*), cholinergic motor neurons positive for *pax6*, *nk6* and *hb9* (*violet [black in print version]*), interneurons positive for *dbx* (*pink [light gray in print version]*), as well as trunk sensory cells positive for *atonal* and *msh* (*light blue [white in print version]*). From Arendt, D., Denes, A.S., Jékely, G., Tessmar-Raible, K., 2008. The evolution of nervous system centralization. Phil. Trans. R. Soc. Lond. B 363, 1523–1528.

In favor of this hypothesis seems to witness studies on the extant lower invertebrates, which suggest that the nerve net-like organization of the nervous system is the basal condition of *Xenacoelomorpha*, including *Xenoturbella* (Gavilán et al., 2016) and the centralization of the nervous system in these groups was an evolutionary process of gradual increase (not decrease!) in the concentration of neurons, which in turn suggests that a similar process of gradual increase in the concentration in the anterior part of the body took place in *Xenacoelomorpha*-grade and Nephrozoa-grade metazoans (Cannon et al., 2016) of the end-Ediacaran to early Cambrian evolution.

According to R.G. Northcutt, Ediacaran biota had only diffuse nerve plexuses and so did the LCBA (last common bilaterian ancestor), which in the course of evolution coalesced into cephalic ganglia and nerve cords or neural tube. The latter increase in the number and cellular differentiation of these structures led to the level of neural complexity that defines them as brains in arthropods, annelids and some molluscs (Northcutt, 2012).

Centralization of the nervous systems was an evolutionary process rather than an "All-or-None" event. Evolution of the nervous system in

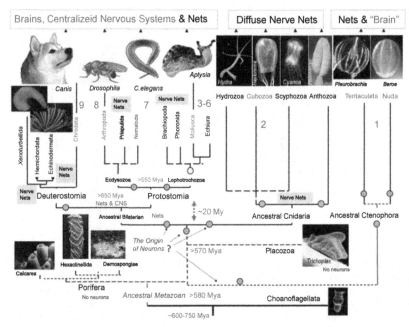

FIGURE 4.37 Parallel evolution of neuronal centralization in the animal kingdom. The diagram summarizes the current view of evolutionary relationships in the animal kingdom and indicates the presence or absence of a central nervous system (CNS) or brain. From this tree it is possible to see at least nine possible events of multiple origins of complex brains — shown as red numbers. Circles indicate possible events of multiple origins of neurons. Possible timing of the divergence in the diagram is indicated as mya. *From Moroz, L., 2012. Phylogenomics meets neuroscience: how many times might complex brains have evolved? Acta Biol. Hung. 63(Suppl. 2), 3—19.*

Xenacoelomorpha, xenoturbellids, and *Acoelomorpha*, represents but "an anterior-posterior gradient" of neurons that "could be interpreted as a first evolutionary step toward nervous system centralization, which is not yet fully expressed in these two basal metazoan clades" (Semmler et al., 2010) The evolution from the diffuse-to the centralized nervous system/brain appears to have occurred through an intermediate ganglionic organization (Northcutt, 2012) of intermediate state or size of anatomic and functional complexity. It is believed that just like the head that evolved from the fusion of several of the anterior body segments, the first brain structures emerged via fusion of several anterior ganglia (Leise et al., 1987; Matheson, 2002), as suggested among other things, by the existence of brain commissures.

Instructive role of the nervous system in development, transgenerational inheritance and speciation

The nervous system in eumetazoans is responsible for their behavior, physiology, and life history. All of them are inextricably linked to each other and to animal life itself. Physiology studies functions of the organism, but if "the function is changed structure", the animal physiology is linked and evolves alongside, and in relation with, animal morphology, for ultimately the function is the raison d'être (reason for being) of the structure. *Herein it is demonstrated that, besides the behavior, physiology, and life history, the nervous system is essentially involved in molding the animal structure and organogenesis.*

The idea of involvement of the nervous system in molding metazoan morphology is not new. Three decades ago, B. John and G. Miklos (1988) intuitively pointed out that "There is every indication that, not only did the evolution of the nervous system involve a far more significant sequence of events than the morphological systems with which most evolutionists have been preoccupied, but, additionally, that its evolution in turn directed many of the morphological changes that have occurred" (John and Miklos, 1988). But B.K. Brian was the first to notice and emphasize the inductive role of the incipient CNS in the post-phylotypic development by giving rise to a network of inductions cell differentiation, histogenesis and organogenesis (Hall, 1998a; Hall, 1998b). Recently G.E. Budd expressed the idea that: "The origins and diversification of the animals, a series of events that became manifest in the so-called 'Cambrian explosion' of *ca* 540 Ma, must necessarily be intimately tied into the evolution of their important organ systems. Of these, the nervous system must be considered to be of extreme importance, not only because of its universality among animals apart from sponges and placozoans, but also because of the role it plays in coordination, sensing and indeed many other aspects of the life of an animal." (Budd, 2015). There is adequate evidence to firmly assert that in metazoans the nervous system/CNS in bilaterians (and cnidarians) took over the control of not only the behavior and physiology but also the building and maintaining the animal structure and morphology (Cabej, 2018b).

This function implies sending instructions (=information) on what to do to the target cells, tissues and organs. By integrating and processing external and internal stimuli, the brain generates this epigenetic information in the form of electrical signals, which within the nervous system are translated into chemical signals to start signal cascades to express/suppress particular genes and induce cell differentiation and organogenesis (Cabej, 2012b).

If the CNS indeed played a role in the Cambrian explosion, that role must be conserved to some degree in the development of the extant taxa, for the evolutionary changes that happened during the Cambrian, ultimately derive from changes that occurred in the developmental pathways of ancestors of the extant taxa. Since developmental pathways are generally conserved across metazoans there is reason to believe that developmental mechanisms of extant animals give us critical clues on the drivers of the developmental changes that brought about the Cambrian explosion. Besides, there is adequate observational evidence on the role of the CNS in transgenerational inheritance of acquired traits, which can be logically extrapolated for explaining mechanisms of evolutionary changes during the Cambrian. The mechanism of the evolutionary change observed in the extant metazoans cannot be different from the one that was operational during the Cambrian. It cannot be a late invention of evolution, for evolution is parsimonious and it would not bother to invent a new mechanism of evolutionary change, different from the one it used during the Cambrian. As an efficient mechanism of induction of evolutionary changes, it would be promoted and conserved by the natural selection. The same can be said for the role of neurocognitive mechanisms in reproductive isolation of extant species in sympatry.

In the following, a part of the uncontested evidence on the instructive role of the nervous system in molding animal morphology and organogenesis, transgenerational inheritance of acquired traits, and neural mechanisms of reproductive isolation, and sympatric speciation will be presented.

A brief record of evidence on inductive role of the CNS in development, organogenesis, inheritance of acquired traits, and speciation

Although no experiments specially designed to investigate the role of the CNS in molding animal morphology have been ever performed, such evidence abounds.

Neural control of development and organogenesis
In insects

- Muscle patterning in *Drosophila* (Lawrence and Johnston, 1986).
- *Drosophila* myogenesis (Currie and Bate, 1991).
- Development of the male-specific muscle or muscle of Lawrence in *Drosophila* (Currie and Bate, 1995).
- Accumulation and proliferation of adult leg muscle precursors in *Manduca sexta* (Consoulas and Levine, 1997).
- Development of flight muscles in *Drosophila* (Fernandes and Keshishian, 1998).
- Muscle apoptosis as a result of proprioceptive denervation in the grasshopper *Barytettix humphreysii* (Arbas and Weidner, 1991; Personius and Chapman, 2002).
- Formation of *Drosophila* trachea and the salivary duct (Kerman and Andrew, 2006).
- Control of the life span in *C. elegans* (Libina et al., 2003; Panowski and Dillin, 2009) and *Drosophila* (Broughton et al., 2005).
- Control of the body growth in *Drosophila* and African malaria mosquito, *Anopheles gambiae* (Rulifson et al., 2002; Colombani et al., 2005; Truman, 2006; Krieger et al., 2004).

- Dealation (wing shedding) in the fire ant, *Solenopsis invicta* (Fletcher and Blum, 1981)
- Determination of wing polyphenisms in the ant *Pheidole megacephala* (Sameshima et al., 2004).
- Determination of dealation in eusocial insects (Robinson and Vargo, 1997; Granger et al., 1996).
- The diphenism winged/unwinged ants (Miura, 2005).
- Horn development in insects of genus *Onthophagus* (Emlen et al., 2005).
- Neural induction of development of the chordotonal organ with auditory function in the eared gipsy moth, *Lymantria dispar* (Lewis and Fullard, 1996).
- Metamorphosis in lower invertebrates and insects is a condensed and highly illustrative example of the general control of the development by the CNS (Cabej, 2012c, 2018c).

In vertebrates

- Induction of head placodes (Leger and Brand, 2002; Chatterjee et al., 2010).
- Inner ear patterning (Legen and Brand, 2002).
- Development of the otic cup (Brigande et al., 2000).
- Regulation of airway branching and morphogenesis in mammals and invertebrates (Bower et al., 2014).
- Determination of the pancreatic islet architecture (Borden et al., 2013).
- Formation of the pituitary gland (Takuma et al., 1998; Treier et al., 2001).
- Tubulogenesis of the salivary gland (Nedvetsky et al., 2014)
- Development of the submandibular (salivary) gland (Knox et al., 2010).
- Patterning of the arterial branching (Mukoyama et al., 2002, 2005; Hogan et al., 2004; Li et al., 2013).
- Specification of adenohypophysis in zebrafish (Herzog et al., 2004).
- Regulation of bone formation (Lundberg et al., 2001; Lerner, 2002; Burt-Pichat et al., 2005).
- Nerve patterning in *X. laevis* (Herrera-Rincon et al., 2017).
- Determination of sex differences in protandrous hermaphroditic fish (Elofsson et al., 1997).
- Regulation of apoptosis (programmed cell death) in medaka fish *Oryzias latipes* (Sugimoto et al., 2000; Uchida-Oka and Sugimoto, 2001).
- Suggested role of pulmonary neuroendocrine cells in lung morphogenesis (Pan et al., 2006; Cutz et al., 2008).
- Differentiation and proliferation of the developing and adult heart (Nebigil et al., 2000).
- Nephrogenesis (Sariola et al., 1988, 1989a, 1989b).

- Temporal regulation of the morphophysiological changes of puberty in vertebrates (Smith et al., 2006; Vidal et al., 2004; Riboni et al., 1998).
- Teeth development (Tuisku and Hildebrand, 1994; Hildebrand et al., 1995; Luukko et al., 1997; Fried et al., 2000; Løes et al., 2002; Kettunen et al., 2005; Zhao et al., 2014; Crucke et al., 2015).
- Formation of dermomyotome and the feather-inducing dermis (Olivera-Martinez et al., 2001; Chang et al., 2004).
- Activation of the lens GRN (gene regulatory network) (Reza and Yasuda, 2004a,b; Carl et al., 2002; Wittbrodt, 2002).
- Determination and maintenance of body mass in mammals (Adams et al., 2001; Levin and Dunn-Meynell, 2000; Baeckberg et al., 2003).
- Induction of limb bud and development (Berggren et al., 1999, 2001; McCaffery and Dräger, 1994).
- Formation of vestibule and cochlea in the inner ear (Riccomagno et al., 2005).
- Formation of the optic cup and lens (Kamachi et al., 1998).
- Amphibian metamorphosis (Denver, 1997a,b; Strand, 1999; Cabej, 2018d).

Neurally induced transgenerational inheritance of acquired characters
Experimental and observational uncontested evidence shows that the nervous system is responsible for abrupt emergence and transmission to progeny of acquired characters.

- Upon visually perceiving their predators, or even on detecting their kairomones, the pea aphid, *Acyrthosiphon pisum*, emits an alarm pheromone, (E)-β-farnesene, whose olfactory perception by conspecific insects induces them to increase the proportion of winged offspring (Dixon and Agarwala, 1999; Kunert and Weiser, 2003; Kunert et al., 2005).
- The cotton aphid, *Aphis gossypii* (Glover), under normal natural conditions produces offspring of four phenotypes: normal green apterae (unwinged), dark green apterae, 'dwarf' yellow apterae, and alatae, but when they detect search track cues of their predator lady bird beetles, *Hippodamia convergens*, they increase the proportion of winged offspring for several sequential generations (Mondor et al., 2005).
- Upon seeing or perceiving kairomones of its predator in the environment *Daphnia cucullata* develops a helmet and doubles the carapace's thickness (Agrawal et al., 1999; Laforsch and Tollrian, 2004).
- Under unfavorable conditions like crowding and shortening of the photoperiod, the small crustacean *Daphnia magna* produces offspring of male sex only. It activates its cholinergic system (Eads et al., 2008) and inhibits the synthesis of neurohormones MOIH-1 (mandibular organ-inhibiting hormone 1) and MOIH-2 by the secretory neurons of the X organ/sinus gland complex, leading to secretion of methyl farnesoate (MF) (Miyakawa et al., 2014) by the mandibular organ. MF forms a complex which activates its receptor MFR (Medlock Kakaley et al., 2017) during a critical period of

oocyte maturation (Rider et al., 2005) and its responsive male sex genes, determining the male sex of the progeny.

- Development of 'neckteeth' by the crustacean *Daphnia pulex* upon detecting the presence in the water of a kairomone released by its predator, the phantom midge larvae *Chaoborus* (the 'neckteeth' prevents the crustacean of being eaten by its predator) (Agrawal et al., 1999; Barry, 2002; Miyakawa et al., 2010; Weiss et al., 2015).

- Adaptive induction of sexuality or diapause by females of several species of *Brachionus* rotifers to their offspring in response to the increase of the concentration of a kairomone by conspecific rotifers, perceived as an indicator of overcrowding (Gilbert, 2004).

- In response to deteriorating conditions (scarce food, crowding, elevated temperature, etc.), the nematode worm *C. elegans* switches to a long life mode (Lee and Ashrafi, 2008) involving changes in expression of 27% of its genes, histone modification (Greer et al., 2011), and non-coding RNAs, which is inherited for at least three generations (Rechavi et al., 2014).

- Induction and inheritance to the offspring of phase transition in two locust species, *Schistocerca gregaria* and *Locusta migratoria*, in response to overcrowding is causally related to changes in the levels of neuroactive substances in the CNS (Rogers et al., 2004; Claeys et al., 2006; Tanaka, 2007; Anstey et al., 2009; Badisco et al., 2011).

Neurally determined reproductive isolation in sympatry and sympatric speciation

- Olfactory and auditory determined sympatric reproductive isolation in the Mexican pupfish (Strecker and Kodric-Brown, 1999) and in *Drosophila* (Ortiz-Barrientos et al., 2004).

- Acoustically determined sympatric speciation of *Gryllus texensis* and *Gryllus rubens* (Gray and Cade, 2000).

- Electrogenic maintenance of reproductive isolation of a group of five morphologically undistinguishable fish species of the genus *Campylomormyrus* in a Central African river basin (Hopkins, 1999) and several morphologically indistinguishable sympatric species (Arnegard et al., 2005; Arnegard et al., 2006).

- Neurocognitive speciation of the apple fruit fly, *Rhagoletis pomonella* (Linn et al., 2003, 2004).

- Two experimentally interbreeding races of the larch budmoth, *Zeiraphera diniana* (Lepidoptera: Tortricidae), do not interbreed in nature based on perception of their own race's mating signals and pheromones released by their females (Emelianov et al., 2003).

- Formation within last centuries of two sympatric species in the American continent from the European corn borer, *Ostrinia nubilalis* (Lepidoptera: Crambidae) by virtue of differences in male preferences for pheromones secreted by females and differences in the time of emergence between the two species (Bush, 1975; Thomas et al., 2003).

- A monophyletic group of forest dwelling *Laupala* crickets in Hawaii evolved with a speciation rate that was 26 times higher that the average speciation rate, based on evolution of female preference to mate conspecific males with courtship song's pulse rate characteristic for each species (Mendelson and Shaw, 2005).
- Mate choice, a neurally determined character, based on the recognition of, and preference for, species-specific visual (Seehausen, 1997a,b; Seehausen and van Alphen, 1998), olfactory (Plenderleith et al., 2005) and auditory (Amorim et al., 2004) mate signals, has been the basic mechanism of evolution and maintenance in sympatry of about 500 sympatric species in Lake Victoria (Africa).
- Rapid speciation of 35 species of Plethodon salamanders in Northeastern United States during the last 5 million years and maintenance of reproductive isolation in sympatry based on the avoidance of interbreeding as a result of conspecific mate preferences (Wiens et al., 2006).

Numerous examples of the intragenerational developmental plasticity also show that neural signals determine changes in developmental pathways, resulting in emergence of new traits. Although not transmitted to the offspring, they corroborate the role of the nervous system in adaptive induction of developmental pathways.

References

Achatz, J.G., Chiodin, M., Salvenmoser, W., Tyler, S., Martinez, P., 2013. The Acoela: on their kind and kinships, especially with nemertodermatids and xenoturbellids (Bilateria incertae sedis). Org. Divers. Evol. 13, 267—286.

Achatz, J., Martínez, P., 2012. The nervous system of Isodiametra pulchra (Acoela) with a discussion on the neuroanatomy of the *Xenacoelomorpha* and its evolutionary implications. Front. Zool. 9, 27.

Adams, C.S., Korytko, A.I., Blank, J.L., 2001. A novel mechanism of body mass regulation. J. Exp. Biol. 204, 1729—1734.

Agata, K., Nakajima, E., Funayama, N., Shibata, N., Saito, Y., Umesono, Y., 2006. Two different evolutionary origins of stem cell systems and their molecular basis. Semin. Cell Dev. Biol. 17, 503—509.

Agata, K., Soejima, Y., Kato, K., Kobayashi, C., Umesono, Y., Watanabe, K., 1998. Structure of the planarian central nervous system (CNS) revealed by neuronal cell markers. Zool. Sci. 15, 433—440.

Agrawal, A.A., Laforsch, C., Tollrian, R., 1999. Transgenerational induction of defences in animals and plants. Nature 401, 60—63.

Amorim, M.C.P., Knight, M.E., Stratoudakis, Y., Turner, G.F., 2004. Differences in sounds made by courting males of three closely related Lake Malawi cichlid species. J. Fish Biol. 65, 1358—1371.

Anstey, M.L., Rogers, S.M., Ott, S.R., Burrows, M., Simpson, S.J., 2009. Serotonin mediates behavioral Gregarization underlying swarm formation in desert locusts. Science 323, 627–630.

Aoki, R., Wake, H., Sasaki, H., Agata, K., 2009. Recording and spectrum analysis of the planarian electroencephalogram. Neuroscience 159, 908–914.

Arbas, E.A., Weidner, M.H., 1991. Transneuronal induction of muscle atrophy in grasshoppers. J. Neurobiol. 22, 536–546.

Arendt, D., Denes, A.S., Jékely, G., Tessmar-Raible, K., 2008. The evolution of nervous system centralization. Phil. Trans. R. Soc. Lond. B 363, 1523–1528.

Arnegard, M.E., Bogdanowicz, S.M., Hopkins, C.D., 2005. Multiple cases of striking genetic similarity between alternative electric fish signal morphs in sympatry. Evolution 59, 324–343.

Arnegard, M.E., Jackson, B.S., Hopkins, C.D., 2006. Time-domain signal divergence and discrimination without receptor modification in sympatric morphs of electric fishes. J. Exp. Biol. 209, 2182–2198.

Ayala, F.J., 1997. Vagaries of the molecular clock. Proc. Natl. Acad. Sci. U.S.A. 94, 7776–7783.

Ayala, F.J., Rzhetsky, A., Ayala, F.J., 1998. Origin of the metazoan phyla: molecular clocks confirm paleontological estimates. Proc. Natl. Acad. Sci. U.S.A. 95, 606–611.

Badisco, L., Huybrechts, J., Simonet, G., Verlinden, H., Marchal, E., Huybrechts, R., et al., 2011. Transcriptome analysis of the desert locust central nervous system: production and annotation of a *Schistocerca gregaria* EST database. PLoS One 6 (3).

Baeckberg, M., Collin, M., Ovesjoe, M.L., Meister, B., 2003. Chemical coding of GABAB receptorimmunoreactive neurons in hypothalamic regions regulating body weight. J. Neuroendocrinol. 15, 1–14.

Baguñà, J., Saló, E., Romero, R., 1989. Effects of activators and antagonists of the neuropeptide substance P and substance K on cell proliferation in planarians. Int. J. Dev. Biol. 33, 261–266.

Bailly, X., Reichert, H., Hartenstein, V., 2013. The urbilaterian brain revisited: novel insights into old questions from new flatworm clades. Dev. Genes Evol. 223, 149–157.

Barry, M.J., 2002. Progress toward understanding the neurophysiological basis of predator-induced morphology in *Daphnia pulex*. Physiol. Biochem. Zool. 75, 179–186.

Bengtson, S., 2001. Biomineralized skeletons — when, where, and why did they evolve? Am. Zool. 41, 1388.

Bengtson, S., 2004. In: Lipps, J.H., Waggoner, B.M. (Eds.), Early Skeletal Fossils. The Paleon- tological Society Papers: Neoproterozoic — Cambrian Biological Revolutions, vol. 10, pp. 67–78.

Bery, A., Martinez, P., 2011. Acetylcholinesterase activity in the developing and regenerating nervous system of the acoel *Symsagittifera roscoffensis*. Acta Zool. 92, 383–392.

Berggren, K., Ezerman, E.B., McCaffery, P., Forehand, C.J., 2001. Expression and regulation of the retinoic acid synthetic enzyme RALDH-2 in the embryonic chicken wing. Dev. Dyn. 222, 1–16.

Berggren, K., McCaffery, P., Dräger, U., Forehand, C.J., 1999. Differential distribution of retinoic acid synthesis in the chicken embryo as determined by immunolocalization of the retinoic acid synthetic enzyme, RALDH-2. Dev. Biol. 210, 288–304.

Bery, A., Cardona, A., Martinez, P., Hartenstein, V., 2010. Structure of the central nervous system of a juvenile acoel, *Symsagittifera roscoffensis*. Dev. Genes Evol. 220, 61–76.

Best, J.B., Goodman, A.B., Pigon, A., 1969. Fissioning in planarians: control by the brain. Science 164, 565–5660.

Boero, F., Gravili, C., Pagliara, P., Piraino, S., Bouillon, J., Schmid, V., 1998. The cnidarian premises of metazoan evolution: from triploblasty, to coelom formation, to metamery. Ital. J. Zool. 65, 5—9.

Boero, F., Schierwater, B., Piraino, S., 2007. Cnidarian milestones in metazoan evolution. Integr. Comp. Biol. 47, 693—700.

Borden, P., Houtz, J., Leach, S.D., Kuruvilla, R., 2013. Sympathetic innervation during development is necessary for pancreatic islet architecture and functional maturation. Cell Rep. 4, 287—301.

Bower, D.V., Lee, H.-K., Lansford, R., Zinn, K., Warburton, D., Fraser, S.E., Jesudason, E.C., 2014. Airway branching has conserved needs for local parasympathetic innervation but not neurotransmission. BMC Biol. 12, 92.

Brigande, J.V., Kiernan, A.E., Gao, X., Iten, L.E., Fekete, D.M., 2000. Molecular genetics of pattern formation in the inner ear: do compartment boundaries play a role? Proc. Natl. Acad. Sci. U.S.A. 97, 11700—11706.

Broughton, S.J., Piper, M.D.W., Ikeya, T., Bass, T.M., Jacobson, J., Driege, Y., et al., 2005. Longer lifespan, altered metabolism, and stress resistance in Drosophila from ablation of cells making insulin-like ligands. Proc. Natl. Acad. Sci. U.S.A. 102, 3105—3110.

Brown, D.D.R., Pearson, B.J., 2017. A brain unfixed: unlimited neurogenesis and regeneration of the adult planarian nervous system. Front. Neurosci. 11, 289.

Brusca, R.C., Moore, W., Shuster, S.M., 2016. In Invertebrates. Chapter 9: Introduction to the Bilateria and the Phylum Xenacoelomorpha — Triploblasty and Bilateral Symmetry Provide New Avenues for Anomal Radiation, Third ed. Sinauer Associates, Inc., Sunderland, MA, U.S.A, p. 354.

Buatois, L.A., 2018. *Treptichnus pedum* and the Ediacaran—Cambrian boundary: significance and caveats. Geol. Mag. 155, 174—180.

Buatois, L.A., Narbonne, G.M., Mángano, M.G., Carmona, N.B., Myrow, P., 2014. Ediacaran matground ecology persisted into the earliest Cambrian. Nat. Commun. 5. Article number: 3544.

Buatois, L.A., Almond, J., Mángano, M.G., Jensen, S., Germs, G.J.B., 2018. Sediment disturbance by Ediacaran bulldozers and the roots of the Cambrian explosion. Sci. Rep. 8, 4514.

Budd, G.E., 2001. Why are arthropods segmented? Evol. Dev. 3 (5), 332—342.

Budd, G.E., 2013. At the origin of animals: the revolutionary cambrian fossil record. Curr. Genomics 14, 344—354.

Budd, G.E., 2015. Early animal evolution and the origins of nervous systems. Philos. Trans. R. Soc. Lond. B Biol. Sci. 370 (1684), 20150037.

Budd, G.E., Jensen, S., 2017. The origin of the animals and a 'Savannah' hypothesis for early bilaterian evolution. Biol. Rev. 92, 446—473.

Bürglin, T.R., Affolter, M., 2016. Homeodomain proteins: an update. Chromosoma 125, 497—521.

Burt-Pichat, B., Lafage-Proust, M.H., Duboeuf, F., Laroche, N., Itzstein, C., Vico, L., et al., 2005. Dramatic decrease of innervation density in bone after ovariectomy. Endocrinology 146, 503—510.

Bush, G.L., 1975. Modes of animal speciation. Annu. Rev. Ecol. Syst. 6, 339—364.

Butterfield, N.J., 2006. Hooking some stem-group 'worms': fossil lophotrochozoans in the Burgess Shale. Bioessays 28, 1161—1166.

Butterfield, N.J., 2007. Macroevolution Macroecol. Deep Palaeontol. 50, 41—55.

Cabej, N.R., 2012b. Epigenetic Principles of Evolution. Elsevier, London — Waltham, MA, pp. 51—61.

Cabej, N.R., 2012c. Ibid., pp. 149—153.

Cabej, N.R., 2018a. Epigenetic Principles of Evolution, Second ed. Academic Press, London — San-Diego, Cambridge MA, Oxford, pp. 146—148.

Cabej, N.R., 2018b. Epigenetic Principles of Evolution, Second edvols. 3—39. Academic Press, London — San-Diego, Cambridge MA, Oxford, pp. 137—214.

Cabej, N.R., 2018c. Ibid, vols. 3—39, pp. 139—144.

Cabej, N.R., 2018d. Ibid, pp. 145—146.

Cannon, J.T., Vellutini, B.C., Smith, J., Ronquist, F., Jondelius, U., Hejnol, A., 2016. *Xenacoelomorpha* is the sister group to Nephrozoa. Nature 530, 89—93.

Carl, M., Loosli, F., Wittbrodt, J., 2002. Six3 inactivation reveals its essential role for the formation and patterning of the vertebrate eye. Development 129, 4057—4063.

Cebrià, F., 2016. Planarian body-wall muscle: regeneration and function beyond a simple skeletal support. Front. Cell Dev. Biol. 4, 8.

Cebrià, F., Newmark, 2007. Morphogenesis defects are associated with abnormal nervous system regeneration following roboA RNAi in planarians. Development 134, 833—837.

Chang, C.H., Jiang, T.X., Lin, C.M., Burrus, L.W., Chuong, C.M., Widelitz, R., 2004. Distinct Wnt members regulate the hierarchical morphogenesis of skin regions (spinal tract) and individual feathers. Mech. Dev. 121, 157—171.

Chatterjee, S., Kraus, P., Lufkin, T., 2010. A symphony of inner ear developmental control genes. BMC Genet. 11, 68.

Chitnis, A., Henrique, D., Lewis, J., Ish-Horowicz, D., Kintner, C., 1995. Primary neurogenesis in *Xenopus* embryos regulated by a homologue of the *Drosophila* neurogenic gene *Delta*. Nature 375, 761—766.

Chen, J.-Y., Hang, D.-Y., Li, C.W., 1999. An early Cambrian craniate-like chordate. Nature 402, 518—522.

Chen, J.-Y., Huang, D.-Y., Peng, Q.-Q., Chi, H.-M., Wang, X.-Q., Feng, M., 2003. The first tunicate from the early cambrian of South China. Proc. Natl. Acad. Sci. U.S.A. 100, 8314—8318.

Chen, J., Li, C., 1997. Early cambrian chordate from Chengjiang, China. Bullet. Natl. Museum Nat. Sci. 10, 257—273.

Chipman, A.D., 2009. Parallel evolution of segmentation by co-option of ancestral gene regulatory networks. Bioessays 32, 60—70.

Chklovskii, D.B., 2004. Synaptic connectivity and neuronal morphology: two sides of the same coin. Neuron 43, 609—617.

Claeys, I., Breugelmans, B., Simonet, G., Van Soest, S., Sas, F., De Loof, A., Vanden Broeck, J., 2006. Neuroparsin transcriptsas molecular markers in the process of desert locust (*Schistocerca gregaria*) phase transition. Biochem. Biophys. Res. Commun. 341, 599—606.

Collins III, J.J., Hou, X., Romanova, E.V., Lambrus, B.G., Miller, C.M., et al., 2015. Correction: genome-wide analyses reveal a role for peptide hormones in planarian germline development. PLoS Biol. 13 (8), e1002234.

Colombani, J., Bianchini, L., Layalle, S., Pondeville, E., Dauphin-Villemant, C., Antoniewski, C., Carre, C., Noselli, S., Leopold, P., 2005. Antagonistic actions of ecdysone and insulins determine final size in *Drosophila*. Science 310, 667—670.

Cong, P., Daley, A.C., Edgecombe, G.D., Hou, X., Chen, A., 2016. Morphology of the radiodontan *Lyrarapax* from the early Cambrian Chengjiang biota. J. Paleontol. 90, 663—671.

Consoulas, C., Levine, R.B., 1997. Accumulation and proliferation of adult leg muscle precursors in *Manduca* are dependent on innervation. J. Neurobiol. 32, 531—553.

Conway Morris, S., 1998. The Crucible of Creation. Oxford University Press, Oxford.

Conway Morris, S., 2000. The Cambrian "explosion": slow-fuse or megatonnage? Proc. Natl. Acad. Sci. U.S.A. 97, 4426–4429.

Conway Morris, S., 2008. A redescription of a rare chordate, metaspriggina walcotti simonetta and insom, from the Burgess Shale (middle cambrian), British Columbia, Canada. J. Pleontol. 82, 424–430.

Conway Morris, S., Caron, J.-B., 2012. Pikaia gracilens walcott, a stem-group chordate from the middle cambrian of British Columbia. Biol. Rev. Camb. Philos. Soc. 87, 480–512.

Conway Morris, S., Peel, J.S., 2008. The earliest annelids: lower cambrian polychaetes from the Sirius Passet Lagerstätte, peary land, north Greenland. Acta Palaeontol. Pol. 53, 137–148.

Conway Morris, S., Whittington, H.B., 1979. The animals of the Burgess Shale. Sci. Am. 122–135.

Conzelmann, M., Williams, E.A., Krug, K., Franz-Wachtel, M., Macek, B., Jékely, G., 2013. The neuropeptide complement of the marine annelid *Platynereis dumerilii*. BMC Genomics 14, 906.

Cossu, G., Borello, U., 1999. Wnt signaling and the activation of myogenesis in mammals. EMBO J. 18, 6867–6872.

Crucke, J., Van de Kelft, A., Huysseune, A., 2015. The innervation of the zebrafish pharyngeal jaws and teeth. J. Anat. 227, 62–71.

Crezée, M., 1975. Monograph of the Solenofilomorphidae (Turbellaria: Acoela). Internationale Revue der gesamten Hydrobiologie und Hydrographie 60, 769–845.

Crezée, M., 1978. *Paratomella rubra* Rieger and Ott, an amphiatlantic acoel turbellarian. Cah. Biol. Mar. 19, 1–9.

Currie, D.A., Bate, M., 1991. The development of adult abdominal muscles in Drosophila: myoblasts express twist and are associated with nerves. Development 113, 91–102.

Currie, D.A., Bate, M., 1995. Innervation is essential for the development and differentiation of a sex-specific adult muscle in *Drosophila melanogaster*. Development 121, 2549–2557.

Cutz, E., Yeger, H., Pan, J., Ito, T., 2008. Pulmonary neuroendocrine cell system in health and disease. Curr. Respir. Med. Rev. 4, 174–186.

Daley, A.C., Antcliffe, J.B., Drage, H.B., Pates, S., 2018. Early fossil record of Euarthropoda and the cambrian explosion. Proc. Natl. Acad. Sci. U.S.A. 115, 5323–5331.

De Robertis, E.M., 2008. Evo-Devo: variations on ancestral themes. Cell 132, 185–195.

Denes, A.S., Jekely, G., Steinmetz, P.R., Raible, F., Snyman, H., Prud'homme, B., Ferrier, D.E., Balavoine, G., Arendt, D., 2007. Molecular architecture of annelid nerve cord supports common origin of nervous system centralization in Bilateria. Cell 129, 277–288.

Denver, R.J., 1997a. Environmental stress as a developmental cue: corticotropin-releasing hormone is a proximate mediator. Horm. Behav. 31, 169–179.

Denver, R.J., 1997b. Thyroid hormone-dependent gene espression program for *Xenopus* neural development. J. Biol. Chem. 272, 8179–8188.

Dewel, R.A., Nelson, D.R., Dewel, W.C., 1993. Tardigrada. In: Harrison, F.W., Rice, M.E. (Eds.), Microscopic Anatomy of Invertebrates. Wiley-Liss, New York, pp. 143–183.

Dixon, A.F.G., Agarwala, B.K., 1999. Ladybird-induced life-history changes in aphids. Proc. R. Soc. Lond. B 266, 1549–1553.

Dolgin, E., 2017. The greatest hits of the human genome. Nature 551, 427–431.

Dunn, C.W., Giribet, G., Edgecombe, G.D., Hejnol, A., 2014. Animal phylogeny and its evolutionary implications. Annu. Rev. Ecol. Evol. Syst. 45, 371–395.

Dupre, C., Yuste, R., 2017. Non-overlapping neural networks in *Hydra vulgaris*. Curr. Biol. 27, 1085–1097.

Dzik, J., 2004. Anatomy and relationships of the early cambrian worm Myoscolex. Zool. Scripta 33, 57–69.

Eads, B.D., Andrews, J., Colbourne, J.K., 2008. Ecological genomics in Daphnia: stress responses and environmental sex determination. Heredity 100, 184–190.

Eldredge, N., Gould, S.J., 1972. Punctuated equilibria: an alternative to phyletic adualism. In: Schopf, T.J.M. (Ed.), Models in Paleobiology. Freeman, San Francisco, pp. 82–115.

Elofsson, U., Winberg, S., Francis, R.C., 1997. Sex differences in number of preoptic GnRH immunoreactive neurons in a protandrously hermaphroditic fish, the anemone fish *Amphiprion melanopus*. J. Comp. Physiol. A 181, 484–492.

Emelianov, I., Simpson, F., Narang, P., Mallet, J., 2003. Host choice promotes reproductive isolation between host races of the larch budmoth *Zeiraphera diniana*. J. Evol. Biol. 16, 208–218.

Emes, R.D., Grant, S.G.N., 2012. Evolution of synapse complexity and diversity. Annu. Rev. Neurosci. 35, 111–131.

Emes, R.D., Pocklington, A.J., Anderson, C.N., Bayes, A., Collins, M.O., Vickers, C.A., et al., 2008. Evolutionary expansion and anatomical specialization of synapse proteome complexity. Nat. Neurosci. 11, 799–806.

Emlen, D.J., Hunt, J., Simmons, L.W., 2005. Evolution of sexual dimorphism in the expression of beetle horns: phylogenetic evidence for modularity, evolutionary lability, and constraint. Am. Nat. 166 (Suppl), S42–S66.

Emmons-Bell, M., Durant, F., Hammelman, J., Bessonov, N., Volpert, V., Morokuma, J., et al., 2015. Species-specific head anatomies in genetically wild-type *Girardia dorotocephala* flatworms. Int. J. Mol. Sci. 16, 27865–27896.

Erwin, D., 2009. Early origin of the bilaterian developmental toolkit. Philos. Trans. R. Soc. Lond. B Biol. Sci. 364, 2253–2261.

Erwin, D.H., Davidson, E.H., 2002. The last common bilaterian ancestor. Development 129, 3021–3032.

Ezkurdia, I., Juan, D., Rodriguez, J.M., Frankish, A., Diekhans, M., Harrow, J., et al., 2014. Multiple evidence strands suggest that there may be as few as 19,000 human protein-coding genes. Hum. Mol. Genet. 23, 5866–5878.

Fan, C.M., Tessier-Lavigne, M., 1994. Patterning of mammalian somites by surface ectoderm and notochord: evidence for sclerotome induction by a hedgehog homolog. Cell 79, 1175–1186.

Fernandes, J.J., Keshishian, H., 1998. Nerve-muscle interactions during flight muscle development in *Drosophila*. Development 125, 1769–1779.

Finnerty, J.R., 2001. Cnidarians reveal intermediate stages in the evolution of hox clusters and axial complexity. Integr. Comp. Biol. 41, 608–620.

Fletcher, D., Blum, M., 1981. Pheromonal control of dealation and oogenesis in virgin queen ants. Science 212, 73–75.

Fortey, R.A., Briggs, D.E.G., Wills, M.A., 1996. The Cambrian evolutionary 'explosion': decoupling cladogenesis from morphological disparity. Biol. J. Linn. Soc. 57, 13–33.

Fraguas, S., Barberan, S., Iglesias, M., Rodriguez-Esteban, G., Cebria, F., 2014. egr-4, a target of EGFR signaling, is required for the formation of the brain primordia and head regeneration in planarians. Development 141, 1835–1847.

Fried, K., Nosrat, C., Lillesaar, C., Hildebrand, C., 2000. Molecular signaling and pulpal nerve development. Crit. Rev. Oral Biol. Med. 11, 318–332 (Abstract).

Friedländer, M.R., Adamidi, C., Han, T., Lebedeva, S., Isenbarger, T.A., Hirst, M., et al., 2009. High-resolution profiling and discovery of planarian small RNAs). Proc. Natl. Acad. Sci. U.S.A. 106, 11546–11551.

Gavilán, B., Perea-Atienza, E., Martínez, P., 2016. Xenacoelomorpha: a case of independent nervous system centralization? Philos. Trans. R. Soc. Lond. B Biol. Sci. 371, 20150039.

Gehling, J.G., 1991. The Case for Ediacaran Fossil Roots to the Metazoan Tree, vol. 20. Geological Society of India Memoir, pp. 181–224.

Genikhovich, G., Technau, U., 2017. On the evolution of bilaterality. Development 144, 3392–3404.

Gilbert, J.J., 2004. Population density, sexual reproduction and diapause in monogonont rotifers: new data for *Brachionus* and a review. J. Limnol. 63 (Suppl. 1), 32–36.

Gould, S.J., 1989. Wonderful Life. Norton, New York.

Graham, A., Butts, T., Lumsden, A., Kiecker, C., 2014. What can vertebrates tell us about segmentation? EvoDevo 5, 24.

Granger, N.A., Sturgis, S.L., Ebersohl, R., Geng, C., Sparks, T.C., 1996. Dopaminergic control of corpora allata activity in the larval tobacco hornworm, *Manduca secta*. Arch. Insect Biochem. 32, 449–466.

Gray, D.A., Cade, W.H., 2000. Sexual selection and speciation in field crickets. Proc. Natl. Acad. Sci. U.S.A. 97, 14449–14454.

Greer, E.L., Maures, T.J., Ucar, D., Hauswirth, A.G., Mancini, E., Lim, J.P., Benayoun, B.A., Shi, Y., Brunet, A., 2011. Transgenerational epigenetic inheritance of longevity in *Caenorhabditis elegans*. Nature 479, 365–371.

Hall, B.K., 1998a. Evolutionary Developmental Biology, second ed. Chapman & Hall, London, p. 163.

Hall, B.K., 1998b. Evolutionary Developmental Biology, second ed. Chapman & Hall, London, p. 134.

Hannibal, R.L., Patel, N.H., 2013. What is a segment? EvoDevo 4, 35.

Hannibal, R.L., Price, A.L., Patel, N.H., 2012. The functional relationship between ectodermal and mesodermal segmentation in the crustacean, *Parhyale hawaiensis*. Dev. Biol. 361, 427–438.

Hejnol, A., 2015. *Acoelomorpha* and Xenoturbellida. In: Wanninger, A. (Ed.), Evolutionary Developmental Biology of Invertebrates 1. Springer, Vienna, pp. 204–214.

Hejnol, A., 2016. Acoelomorpha. In: Schmidt-Rhaesa, A., Harzsch, S., Purschke, G. (Eds.), Structure and Evolution of Invertebrate Nervous Systems. Oxford University Press, Oxford, pp. 56–61.

Hejnol, A., Pang, K., 2016. Xenacoelomorpha's significance for understanding bilaterian evolution. Curr. Opin. Genet. Dev. 39, 48–54.

Hench, J., Henriksson, J., Abou-Zied, A.M., Lüppert, M., Dethlefsen, J., Mukherjee, K., et al., 2015. The Homeobox genes of *Caenorhabditis elegans* and insights into their spatio-temporal expression dynamics during embryogenesis. PLoS One 10 (5), e0126947.

Herculano-Houzel, S., 2012. The remarkable, yet not extraordinary, human brain as a scaled-up primate brain and its associated cost. Proc. Natl. Acad. Sci. U.S.A. 109 (Suppl. 1), 10661–10668.

Herrera-Rincon, C., Pai, V.P., Moran, K.M., Lemire, J.M., Levin, M., 2017. The brain is required for normal muscle and nerve patterning during early *Xenopus* development. Nat. Commun. 8, 587.

Herzog, W., Sonntag, C., von der Hardt, S., Roehl, H.H., Varga, Z.M., Hammerschmidt, M., 2004. Fgf3 signalling from the ventral diencephalon is required for early specification and subsequent survival of the zebrafish adenohypophysis. Development 131, 3681–3692.

Hildebrand, C., Fried, K., Tuisku, F., Johansson, C.S., 1995. Teeth and tooth nerves. Prog. Neurobiol. 45, 165–222.

Hodgkin, J., 2001. What does a worm want with 20,000 genes? Genome Biol. 2 (11) comment 2008).

Hogan, K.A., Ambler, C.A., Chapman, D.L., Bautch, V.L., 2004. The neural tube patterns vessels developmentally using the VEGF signaling pathway. Development 131, 1503–1513.

Holland, N.D., 2003. Early central nervous system evolution: an era of skin brains? Nat. Rev. Neurosci. 4, 617–627.

Holland, L.Z., 2015. The origin and evolution of chordate nervous systems. Phil. Trans. R. Soc. B 370, 20150048.

Holland, L.Z., Carvalho, J.E., Escriva, H., Laudet, V., Schubert, M., Shimeld, S.M., et al., 2013. Evolution of bilaterian central nervous systems: a single origin? EvoDevo 4, 27.

Hopkins, C.D., 1999. Signal evolution in electric communication. In: Hauser, M.D., Konishi, M. (Eds.), The Design of Animal Communication. MIT Press, Cambridge, MA, pp. 461–491.

Inoue, T., Hoshino, H., Yamashita, T., Shimoyama, S., Agata, K., 2015. Planarian shows decision-making behavior in response to multiple stimuli by integrative brain function. Zool. Lett. 1–7.

Inoue, T., Kumamoto, H., Okamoto, K., Umesono, Y., Sakai, M., Alvarado, A.S., Agata, K., 2004. Morphological and functional recovery of the planarian photosensing system during head regeneration. Zool. Sci. 21, 275–283.

Israelsson, O., 2007. Ultrastructural aspects of the 'statocyst' of Xenoturbella (Deuterostomia) cast doubt on its function as a georeceptor. Tissue Cell 39, 171–177.

Jacobs, D.K., Hughes, N.C., Fitz-Gibbon, S.T., Winchel, C.J., 2005. Terminal addition, the Cambrian radiation and thePhanerozoic evolution of bilaterian form. Evol. Dev. 7 (6), 498–514.

Joel, L.V., Droser, M.L., Gehling, J.G., 2014. A new enigmatic, tubular organism from the Ediacara member, Rawnsley quartzite, South Australia. J. Paleontol. 88, 253–262.

John, B., Miklos, G., 1988. The Eukaryote Genome in Development and Evolution. Allen & Unwin, London – Boston, p. 331.

Kamachi, Y., Uchikawa, M., Collignon, J., Lovell-Badge, R., Kondoh, H., 1998. Involvement of Sox1, 2 and 3 in the early and subsequent molecular events of lens induction. Development 125, 2521–2532.

Kerman, B.K., Andrew, D.J., 2006. Analysis of Dalmatian suggests a role for the nervous system in drosophila embryonic trachea and salivary duct development. 47th Drosophila Research Conference Abstracts, p. 275.

Kettunen, P., Løes, S., Furmanek, T., Fjeld, K., Kvinnsland, I.H., Behar, O., Yagi, T., et al., 2005. Coordination of trigeminal axon navigation and patterning with tooth organ formation: epithelial-mesenchymal interactions, and epithelial Wnt4 and Tgfβ1 regulate semaphorin 3a expression in the dental mesenchyme. Development 132, 323–334.

Knoll, A.K., 2004. Life on a Young Planet: The First Three Billion Years of Evolution. Princeton University Press, Princeton, NJ.

Knox, S.M., Lombaert, I.M.A., Reed, X., Vitale-Cross, L., Gutkind, J.S., Hoffman, M.P., 2010. Parasympathetic innervation maintains epithelial progenitor cells during salivary organogenesis. Science 329, 1645–1647.

Krämer-Eis, A., Ferretti, L., Schiffer, P.H., Heger, P., Wiehe, T., 2016. The Common Developmental Genetic Toolkit of Bilaterian Crown Clades after a Billion Years of Divergence. bioRxiv. Available from: https://doi.org/10.1101/041806.

Krieger, M.J.B., Jahan, N., Riehle, M.A., Cao, C., Brown, M.R., 2004. Molecular characterization of insulin-like peptide genes and their expression in the African malaria mosquito, *Anopheles gambiae*. Insect Mol. Biol. 13, 305–315.

Kumar, A., Brocke, J.P., 2012. Nerve dependence in tissue, organ, and appendage regeneration. Trends Neurosci. 35, 691–699.

Kunert, G., Otto, S., Roese, U.S.R., Gershenzon, J., Weisser, W.W., 2005. Alarm pheromone mediates production of winged dispersal morphs in aphids. Ecol. Lett. 8, 596–603.

Kunert, G., Weiser, W.W., 2003. The interplay between density- and trait-mediated effects in predatory-prey interactions: a case study in aphid wing polymorphism. Oecologia 135, 304–312.

Lacalli, T., 2012. The Middle Cambrian fossil Pikaia and the evolution of chordate swimming. EvoDevo 3, 12.

Laforsch, C., Tollrian, R., 2004. Extreme helmet formation in *Daphnia cucullata* induced by smallscale turbulence. J. Plankton Res. 26, 81–87.

Landing, E., Geyer, G., Bartowski, K.E., 2002. Latest early cambrian small shelly fossils, trilobites, and hatch Hill dysaerobic interval on the Quebec continental slope. J. Paleontol. 76, 287–305.

Lawrence, P.A., Johnston, P., 1986. The muscle pattern of a segment of *Drosophila* may be determined by neurons and not by contributing myoblasts. Cell 45, 505–513.

Lee, B.H., Ashrafi, K., 2008. A TRPV channel modulates C. elegans neurosecretion, larval starvation survival, and adult lifespan. PLoS Genetics 4 (10), e1000213. https://doi.org/10.1371/journal.pgen.1000213.

Leger, S., Brand, M., 2002. Fgf8 and Fgf3 are required for zebrafish ear placode induction, maintenance and inner ear patterning. Mech. Dev. 119, 91–108.

Leise, E.M., Hall, W.M., Mulloney, B., 1987. Functional organization of crayfish abdominal ganglia. II. Sensory afferents and extensor motor neurons. J. Comp. Neurol. 266, 495–518.

Levin, B.E., Dunn-Meynell, A.A., 2000. Sibutramine alters the central mechanisms regulating the defended body weight in diet-induced obese rats. Am. J. Physiol. Reg. I. 279, R2222–R2228.

Lewis, F.P., Fullard, J.H., 1996. Neurometamorphosis of the ear in the gypsy moth, *Lymnatria dispar*, and its homologue in the earless forest tent caterpillar moth, *Malacosoma disstria*. J. Neurobiol. 31, 245–262.

Leys, S.P., 2015. Elements of a 'nervous system' in sponges. J. Exp. Biol. 218, 581–591.

Li, W., Kohara, H., Uchida, Y., James, J.M., Soneji, K., Cronshaw, D.G., et al., 2013. Peripheral nerve-derived CXCL12 and VEGF-A regulate the patterning of arterial vessel branching in developing limb skin. Dev. Cell 24, 359–371.

Libina, N., Berman, J.R., Kenyon, C., 2003. Tissue-specific activities of *C. elegans* DAF-16 in the regulation of lifespan. Cell 115, 489–502.

Lieberman, B.S., 2003. Taking the pulse of the cambrian radiation. Integr. Comp. Biol. 43, 229–237.

Lieberman, B.S., 2008. The Cambrian radiation of bilaterians: evolutionary origins and palaeontological emergence; earth history change and biotic factors. Palaeogeogr. Palaeoclimatol. Palaeoecol. 258, 180–188.

Lin, J.-P., Gon III, S.M., Gehling, J.G., Loren, B., Zhao, Y.-L., Zhang, X.-L., et al., 2006. A *Parvancorina*-like arthropod from the cambrian of south China. Hist. Biol. 18, 33–45.

Linnemann, U., Ovtcharova, M., Schaltegger, U., Gärtner, A., Hautmann, M., Geyer, G., et al., 2019. New high-resolution age data from the Ediacaran–Cambrian boundary indicate rapid, ecologically driven onset of the Cambrian explosion. Terra. Nova 31, 49–58.

Linn, C.E., Dambroski, H.R., Feder, J.L., Berlocher, S.H., Nojima, S., Roelofs, W.L., 2004. Postzygotic isolating factor in sympatric speciation in Rhagoletis flies: reduced response of hybrids to parental host-fruit odors. Proc. Natl. Acad. Sci. U.S.A. 101, 17753—17758. List of animals by number of neurons. Available from: https://howlingpixel.com/i-en/List_of_animals_by_number_of_neurons.

Linn, C., Feder, J.L., Nojima, S., Dambroski, H.R., Berlocher, S.H., Roelofs, W., 2003. Fruit odor discrimination and sympatric host race formation in Rhagoletis. Proc. Natl. Acad. Sci. U.S.A. 100, 11490—11493.

Lobo, D., Beane, W.S., Levin, M., 2012. Modeling planarian regeneration: a primer for reverse-engineering the worm. PLoS Comput. Biol. 8 (4), e1002481.

Løes, S., Kettunen, P., Kvinnsland, H., Luukko, K., 2002. Mouse rudimentary diastema tooth primordia are devoid of peripheral nerve fibers. Anat. Embryol. (Berl.). 205, 187—191.

Lowe, C.J., Wu, M., Salic, A., Evans, L., Lander, E., Stange-Thomann, N., et al., 2003. Origins of the chordate nervous system. Cell 113, 853—865.

Lundberg, P., Lundgren, I., Mukohyama, H., Lehenkari, P.P., Horton, M.A., Lerner, U.H., 2001. Vasoactive intestinal peptide (VIP)/pituitary adenylate cyclase-activating peptide receptor subtypes in mouse calvarial osteoblasts: presence of VIP-2 receptors and differentiation-induced expression of VIP-1 receptors. Endocrinology 142, 339—347.

Luukko, K., Sariola, H., Saarma, M., Thesleff, I., 1997. Localization of nerve cells in the developing rat tooth. J. Dent. Res. 76, 1350—1356.

Mángano, M.G., Buatois, L.A., 2016. The Cambrian explosion (Chapter 3). In: Mángano, M.G., Buatois, L.A. (Eds.), The Trace-Fossil Record of Major Evolutionary Events, Topics in Geobiology, vol. 39. Springer, Dordrecht, pp. 77—126, 73.

Mángano, M.G., Buatois, L.A., 2014. Decoupling of body-plan diversification and ecological structuring during the Ediacaran-Cambrian transition: evolutionary and geobiological feedbacks. Proc. R. Soc. Lond. B Biol. Sci. 281 (1780), 20140038.

Marlow, H.Q., Srivastava, M., Matus, D.Q., Rokhsar, D., Martindale, M.Q., 2009. Anatomy and development of the nervous system of Nematostella vectensis, an anthozoan Cnidarian. Dev. Neurobiol. 69, 235—254.

Marshall, C.R., 2006. Explaining the cambrian "explosion" of animals. Annu. Rev. Earth Planet Sci. 34, 355—384.

Martín-Durán, J.M., Monjo, F., Romero, R., 2012. Planarian embryology in the era of comparative developmental biology. Int. J. Dev. Biol. 56, 39—48.

Martín-Durán, J.M., Pang, K., Børve, A., Lê, H.S., Furu, A., Cannon, J.T., Jondelius, U., Hejnol, A., 2018. Convergent evolution of bilaterian nerve cords. Nature 553, 45—50.

Martindale, M.Q., Pang, K., Finnerty, J.R., 2004. Investigating the origins of triploblasty: 'mesodermal' gene expression in a diploblastic animal, the sea anemone Nematostella vectensis (phylum, Cnidaria; class, Anthozoa). Development 131, 2463—2474.

Martinez, P., Perea-Atienza, E., Gavilán, B., Fernandez, C., Sprecher, S., 2017. The study of xenacoelomorph nervous systems. Molecular and morphological perspectives. Invertebr. Zool. 14, 32—44.

Matheson, T., 2002. Invertebrate Nervous Systems. Encyclopedia of Life Sciences. McMillan Publishers Ltd. Available from: https://www2.le.ac.uk/departments/npb/people/matheson/matheson-neurobiology/images/publications/Matheson_ELS_2002.pdf.

McCaffery, P., Dräger, U.C., 1994. Hot spots of retinoic acid synthesis in the developing spinal cord. Proc. Natl. Acad. Sci. U.S.A. 91, 7194—7197.

Medlock Kakaley, E.K., Wang, H.Y., LeBlanc, G.A., 2017. Agonist-mediated assembly of the crustacean methyl farnesoate receptor. Sci. Rep. 7.

Meinhardt, H., 2006. Primary body axes of vertebrates: generation of a near-Cartesian coordinate system and the role of Spemann-type organizer. Dev. Dynam. 235, 2907—2919.

Mendelson, T.C., Shaw, K.L., 2005. Rapid speciation in an arthropod. Nature 433, 375—376.

Mineta, K., Nakazawa, M., Cebrià, F., Ikeo, K., Agata, K., Gojobori, T., 2003. Origin and evolutionary process of the CNS elucidated by comparative genomics analysis of planarian ESTs. Proc. Natl. Acad. Sci. U.S.A. 100, 7666—7671.

Miura, T., 2005. Developmental regulation of caste-specific characters in social-insect polyphenism. Evol. Dev. 7, 122—129.

Miyakawa, H., Imai, M., Sugimoto, N., Ishikawa, Y., Ishikawa, A., Ishigaki, H., Okada, Y., Miyazaki, S., Koshikawa, S., Cornette, R., Miura, T., 2010. Gene up-regulation in response to predator kairomones in the water flea, *Daphnia pulex*. BMC Dev. Biol. 10, 45.

Miyakawa, H., Toyota, K., Sumiya, E., Iguchi, T., 2014. Comparison of JH signaling in insects and crustaceans. Curr. Opin. Insect Sci. 1, 81—87.

Miyamoto, S., Shimozawa, A., 1985. Chemotaxis in the freshwater planarian, *Dugesia japonica japonica*. Zool. Sci. 2, 389—395.

Mondor, E.B., Rosenheim, J.A., Addicott, J.F., 2005. Predator-induced transgenerational phenotypic plasticity in the cotton aphid. Oecologia 142, 104—108.

Monk, T., Paulin, M.G., 2014. Evolutionary origins of sensation in metazoans: functional evidence for a new sensory organ in sponges. Brain Behav. Evol. 84, 246—261.

Moroz, L., 2012. Phylogenomics meets neuroscience: how many times might complex brains have evolved? Acta Biol. Hung. 63 (Suppl. 2), 3—19.

Mouse Genome Sequencing Consortium, 2002. Initial sequencing and comparative analysis of the mouse genome. Nature 420, 520—561.

Mueller, C.T., 2002. The use of classical conditioning in planaria to investigate a non-neuronal memory mechanism. Available from: http://drmichaellevin.org/Planaria/prelimdata/CMueller.pdf.

Mukoyama, Y., Gerber, H.-P., Ferrara, N., Gu, C., Anderson, D.J., 2005. Peripheral nerve-derived VEGF promotes arterial differentiation via neuropilin 1-mediated positive feedback. Development 132, 941—952.

Mukoyama, Y.S., Shin, D., Britsch, S., Taniguchi, M., Anderson, D.J., 2002. Sensory nerves determine the pattern of arterial differentiation and blood vessel branching in the skin. Cell 109, 693—705.

Nakano, H., 2015. What is *Xenoturbella*? Zool. Lett. 1, 22.

Nakazawa, M., Cebrià, F., Mineta, K., Ikeo, K., Agata, K., Gojobori, T., 2003. Search for the evolutionary origin of a brain: planarian brain characterized by microarray. Mol. Biol. Evol. 20, 784—791.

Nanglu, K., Caron, J.-B., Conway Morris, S., Cameron, C.B., 2016. Cambrian suspension-feeding tubicolous hemichordates. BMC Biol. 14, 56.

Nebigil, C.G., Choi, D.-S., Dierich, A., Hickel, P., Le Meur, M., Messaddeq, N., et al., 2000. Serotonin 32B receptor is required for heart development. Proc. Natl. Acad. Sci. U.S.A. 97, 9508—9513.

Nedvetsky, P.I., Emmerson, E., Finley, J.K., Ettinger, A., Cruz-Pacheco, N., Prochazka, J., et al., 2014. Parasympathetic innervation regulates tubulogenesis in the developing salivary gland. Dev. Cell 30, 449—462.

Neuhof, M., Levin, M., Rechavi, O., 2016. Vertically- and horizontally-transmitted memories — the fading boundaries between regeneration and inheritance in planaria. Biology Open 5, 1177—1188.

Newmark, P.A., Alvarado, A.S., 2002. Not your father's planarian: a classic model enters the era of functional genomics. Nat. Rev. Genet. 3, 210–219.

Nichols, S.A., Dirks, W., Pearse, J.S., King, N., 2006. Early evolution of animal cell signaling and adhesion genes. Proc. Natl. Acad. Sci. U.S.A. 103, 12451–12456.

Nicolas, C.L., Abramson, C.I., Levin, M., 2008. Analysis of behavior in the planarian model. In: Raffa, R.B., Rawls, S.M. (Eds.), Planaria: A Model for Drug Action and Abuse. Landes Bioscience, Austin, TX, pp. 83–94.

Nielsen, C., 2012. Animal Evolution: Interrelationships of the Living Phyla, third ed. Oxford University Press, Oxford, UK, p. 348.

Nielsen, C., 2012b. Animal Evolution: Interrelationships of the Living Phyla, third ed. Oxford University Press, Oxford, UK, p. 349.

Nielsen, C., 2012c. Ibid., p. 350.

Nielsen, C., 2013d. Ibid, p. 354.

Northcutt, R.G., 2012. Evolution of centralized nervous systems. Proc. Natl. Acad. Sci. U.S.A. 109 (Suppl. 1), 10626–10633.

Olivera-Martinez, I., Thelu, J., Teillet, M.-A., Dhouailly, D., 2001. Dorsal dermis development depends on a signal from the dorsal neural tube, which can be substituted by Wnt-1. Mech. Dev. 100, 233–244.

Ortiz-Barrientos, D., Counterman, B.A., Noor, M.A., 2004. The genetics of speciation by reinforcement. PLoS Biol. 2, e416.

Oviedo, N.J., Morokuma, J., Walentek, P., Kema, I.P., Gu, M.B., Ahn, J.-M., et al., 2010. Long-range neural and gap junction protein-mediated cues control polarity during planarian regeneration. Dev. Biol. 339, 188–199.

Owlarn, S., Bartscherer, K., 2016. Go ahead, grow a head! A planarian's guide to anterior regeneration. Regeneration (Oxf.) 3, 139–155.

Pan, J., Copland, I., Post, M., Yeger, H., Cutz, E., 2006. Mechanical stretch-induced serotonin release from pulmonary neuroendocrine cells: implications for lung development. Am. J. Physiol. Lung Mol. Physiol. 290, L185–L193.

Panowski, S.H., Dillin, A., 2009. Signals of youth: endocrine regulation of aging in *Caenorhabditis elegans*. Trends Endocrinol. Metab. 20, 259–264.

Parry, L., 2014. Fossil focus: annelids. Palaeontol. 4 (11), 1–8. Online.

Paskin, T.R., Jellies, J., Bacher, J., Beane, W.S., 2014. Planarian phototactic assay reveals differential behavioral responses based on wavelength. PLoS One 9 (12), e114708.

Paterson, J.R., Edgecombe, G.D., Lee, M.S.Y., 2019. Trilobite evolutionary rates constrain the duration of the Cambrian explosion. Proc. Natl. Acad. Sci. U.S.A. 116, 4394–4399.

Perea-Atienza, E., Gavilán, B., Chiodin, M., Abril, J.F., Hoff, K.J., Poustka, A.J., Martinez, P., 2015. The nervous system of Xenacoelomorpha: a genomic perspective. J. Exp. Biol. 218, 618–628.

Personius, K.E., Chapman, R.F., 2002. Control of muscle degeneration following autotomy of a hindleg in the grasshopper, *Barytettix humphreysii*. J. Insect Physiol. 48, 91–102.

Petralia, R.S., Wang, Y.X., Mattson, M.P., Yao, P.J., 2016. The diversity of spine synapses in animals. NeuroMolecular Med. 18, 497.

Plenderleith, M., van Oosterhout, C., Robinson, R.L., Turner, G.F., 2005. Female preference for conspecific males based on olfactory cues in a Lake Malawi cichlid fish. Biol. Lett. 1, 411–414.

Putnam, N.H., Srivastava, M., Hellsten, U., Dirks, B., Chapman, J., Salamov, A., et al., 2007. sea anemone genome reveals ancestral eumetazoan gene repertoire and genomic organization. Science 317, 86–94.

Raikova, O.I., Reuter, M., Gustafsson, M.K., Maule, A.G., Halton, D.W., Jondelius, U., 2004. Basiepidermal nervous system in *Nemertoderma westbladi* (Nemertodermatida): GYIRFamide immunoreactivity. Zoology 107, 75–86.

Raikova, O.I., Reuter, M., Jondelius, U., Gustafsson, M.K.S., 2000. An immunocytochemical and ultrastructural study of the nervous and muscular systems of *Xenoturbella westbladi* (Bilateria inc. sed.). Zoomorphology 120, 107–118.

Rechavi, O., Houri-Ze'evi, L., Anava, S., Goh, W.S.S., Kerk, S.Y., Hannon, G.J., Hobert, O., 2014. Starvation-induced transgenerational inheritance of small RNAs in *C. elegans*. Cell 158, 277–287.

Reddien, P.W., Sánchez Alvarado, A., 2004. Fundamentals of planarian regeneration. Annu. Rev. Cell Dev. Biol. 20, 725–757.

Rentzsch, F., Holstein, T.W., 2018. Making head or tail of cnidarian *hox* gene function. Nat. Commun. 9, 2187.

Reza, H.M., Yasuda, K., 2004a. Lens differentiation and crystallin regulation: a chick model. Int. J. Dev. Biol. 48, 805–817.

Reza, H.M., Yasuda, K., 2004b. The involvement of neural retina Pax6 in lens fiber differentiation. Dev. Neurosci. 26, 318–327.

Riboni, L., Escamilla, C., Chavira, R., Dominguez, R., 1998. Effects of peripheral sympathetic denervation induced by guanethidine administration on the mechanisms regulating puberty in the female Guinea pig. J. Endocrinol. 156, 91–98.

Riccomagno, M.M., Takada, S., Epstein, D.J., 2005. Wnt-dependent regulation of inner ear morphogenesis is balanced by the opposing and supporting roles of Shh. Genes Dev. 19, 1612–1623.

Rider, C.V., Gorr, T.A., Olmstead, A.W., Wasilak, B.A., LeBlanc, G.A., 2005. Stress signaling: coregulation of hemoglobin and male sex determination through a terpenoid signaling pathway in a crustacean. J. Exp. Biol. 208, 15–23.

Robinson, G.E., Vargo, E.L., 1997. Juvenile hormone in adult eusocial Hymenoptera: gonadotropin and behavioral pacemaker. Arch. Insect Biochem. 35, 559–583.

Rogers, S.M., Matheson, T., Sasaki, K., Kendrick, K., Simpson, S.J., Burrows, M., 2004. Substantial changes in central nervous system neurotransmitters and neuromodulators accompany phase change in the locust. J. Exp. Biol. 207, 3603–3617.

Ross, K.G., Currie, K.W., Pearson, B.J., Zayas, R.M., 2017. Nervous system development and regeneration in freshwater planarians. WIREs Dev. Biol. 6, e266.

Rouhana, L., Tasaki, J., Saberi, A., Newmark, P.A., 2017. Genetic dissection of the planarian reproductive system through characterization of *Schmidtea mediterranea* CPEB homologs. Dev. Biol. 426, 43–55.

Rulifson, E.J., Kim, S.K., Nusse, R., 2002. Ablation of insulin-producing neurons in flies: growth and diabetic phenotypes. Science 296, 1118–1120.

Runnegar, B., 1986. Molecular palaeontology. Palaeontology 29, 1–24.

Ryan, J.F., Burton, P.M., Mazza, M.E., Kwong, G.K., Mullikin, J.C., Finnerty, J.R., 2006. The cnidarian-bilaterian ancestor possessed at least 56 homeoboxes: evidence from the starlet sea anemone, *Nematostella vectensis*. Genome Biol. 7, R64.

Saberi, A., Jamal, A., Beets, I., Schoofs, L., Newmark, P.A., 2016. GPCRs direct germline development and somatic gonad function in planarians. PLoS Biol. 14 (5), e1002457.

Sameshima, S.-Y., Miura, T., Matsumoto, T., 2004. Wing disc development during caste differentiation in the ant Pheidole megacephala (Hymenoptera: Formicidae). Evol. Dev. 6, 336–341.

Sariola, H., Ekblom, P., Henke-Fahle, S., 1989a. Embryonic neurons as in vitro inducers of differentiation of nephrogenic mesenchyme. Dev. Biol. 132, 271–281.

Sariola, H., Holm, K., Heinke-Fahle, S., 1988. Early innervation of the metanephric kidney. Development 104, 589–599.

Sariola, H., Holm-Sainio, K., Henke-Fahle, S., 1989b. The effect of neuronal cells on kidney differentiation. Int. J. Dev. Biol. 33, 149–155.

Sarnat, H.B., Netsky, M.G., 1985. The brain of the planarian as the ancestor of the human brain. Can. J. Neurol. Sci. 12, 296–302.

Satoh, N., Rokhsar, D., Nishikawa, T., 2014. Chordate evolution and the three-phylum system. Proc. Biol. Sci. 281, 20141729.

Satterlie, R., 2018. Cnidarian Neurobiology. In: Byrne, J.H. (Ed.), The Oxford Handbook of Invertebrate Neurobiology. Oxford University Press, 2018 Online Publication Date: Feb 2017.

Schmidt-Rhaea, A., 2007. The Evolution of Organ Systems. Oxford University Press, NewYork, p. 34.

Schoenemann, B., Pärnaste, H., Clarkson, E.N.K., 2017. Structure and function of a compound eye, more than half a billion years old. Proc. Natl. Acad. Sci. U.S.A. 114, 13489–13494.

Schopf, J.W., 1992. Major Events in the History of Life. Jones & Bartlett Publishers, p. 76.

Seehausen, O., van Alphen, J.J.M., 1998. The effect of male coloration on female mate choice in closely related Lake Victoria cichlids (*Haplochromis nyererei* complex). Behav. Ecol. Sociobiol. 42, 1–8.

Seehausen, O., van Alphen, J.J.M., Witte, F., 1997a. Cichlid fish diversity threatened by eutrophication that curbs sexual selection. Science 277, 1808–1811.

Seehausen, O., Witte, F., Katunzi, E.F.B., Smits, J., Bouton, N., 1997b. Patternsof the remnant cichlid fauna in southern Lake Victoria. Conserv. Biol. 11, 890–905.

Seilacher, A., Buatois, L.A., Mángano, M.G., 2005. Trace fossils in the Ediacaran-Cambrian transition: behavioral diversification, ecological turnover and environmental shift. Palaeogeogr. Palaeoclimatol. Palaeoecol. 227, 323–356.

Semmler, H., Wanninger, A., Bailly, X., Chiodin, M., Martinez, P., 2010. Steps towards a centralized nervous system in basal bilaterians: insights from neurogenesis of the acoel *Symsagittifera roscoffensis*. Dev. Growth Differ. 52, 701–713.

Shimoyama, S., Inoue, T., Kashima, M., Agata, K., 2016. Multiple neuropeptide-coding genes involved in planarian pharynx extension. Zool. Sci. 33, 311–319.

Shu, D.-G., Conway Morris, S., Zhang, X.-L., 1996b. A pikaia-like chordate from the lower Cambrian of China. Nature 384, 157–158.

Shu, D.-G., Luo, H.-L., Conway Morris, S., Zhang, X.-L., Hu, S.-X., Chen, L., et al., 1999. Lower Cambrian vertebrates from South China. Nature 402, 42–46.

Shu, D., Zhang, X., Chen, L., 1996a. Reinterpretation of *Yunnanozoon* as the earliest known hemichordate. Nature 380, 428–430.

Smith, J.T., Clifton, D.K., Steiner, R.A., 2006. Regulation of the neuroendocrine reproductive axis by kisspeptin-GPR54 signaling. Reproduction 131, 623–630.

Smith, E.F., Nelson, L.L., Strange, M.A., Eyster, A.E., Rowland, S.M., Schrag, D.P., Macdonald, F.A., 2016. The end of the Ediacaran: two new exceptionally preserved body fossil assemblages from Mount Dunfee, Nevada, U.S.A. Geology 44, 911–914.

Smith, S.L., Smith, I.T., Branco, T., Häusser, M., 2013. Dendritic spikes enhance stimulus selectivity in cortical neurons *in vivo*. Nature 503, 115–120.

Smith, J., Tyler, S., 1985. The acoel turbellarians: kingpins of metazoan evolution or a specialized offshoot? In: Conway Morris, S., George, J.D., Gibson, R., Platt, H.M. (Eds.), The Origins and Relationships of Lower Invertebrates. Calderon Press, Oxford, pp. 123–142.

Solana, J., 2013. Closing the circle of germline and stem cells: the primordial stem cell hypothesis. EvoDevo 4, 2.

Srivastava, M., Simakov, O., Chapman, J., Fahey, B., Gauthier, M., et al., 2010. The *Amphimedon queenslandica* genome and the evolution of animal complexity. Nature 466, 720–726.

Steele, R.E., David, C.N., Technau, U., 2011. A genomic view of 500 million years of cnidarian evolution. Trends Genet. 27, 7–13.

Stein, M., Budd, G.E., Peel, J.S., Harper, D.A.T., 2013. *Arthroaspis* n. gen., a common element of the Sirius Passet Lagerstätte (Cambrian, North Greenland), sheds light on trilobite ancestry. BMC Evol. Biol. 13, 99.

Steiner, M., Li, G., Qian, Y., Zhu, M., Erdtmann, B.D., 2007. Neoproterozoic to Early Cambrian small shelly fossil assemblages and a revised biostratigraphic correlation of the Yangtze Platform (China). Palaeogeogr. Palaeoclimatol. Palaeoecol. 254, 67–99.

Strand, F.L., 1999. Neuropeptides. MIT Press, Cambridge, MA, p. 183.

Strausfeld, N.J., Hirth, F., 2013. Deep homology of arthropod central complex and vertebrate basal ganglia. Science 340, 157–161.

Strausfeld, N.J., Ma, X., Edgecombe, G.D., 2016. Fossils and the evolution of the arthropod brain. Curr. Biol. 26, R989–R1000.

Strecker, U., Kodric-Brown, A., 1999. Mate recognition systems in a species flock of Mexican pupfish. J. Evol. Biol. 12, 927–935.

Sugimoto, M., Uchida, N., Hatayama, M., 2000. Apoptosis in skin pigmentcells of medaka, *Oryzias latipes* (Teleostei), during long-term chromatic adaptation: the role of sympathetic innervation. Cell Tissue Res. 301, 205–216 (Abstract).

Swapna, L.S., Molinaro, A.M., Lindsay-Mosher, N., Pearson, B.J., Parkinson, J., 2018. Comparative transcriptomic analyses and single-cell RNA sequencing of the freshwater planarian *Schmidtea mediterranea* identify major cell types and pathway conservation. Genome Biol. 19, 124.

Takeda, H., Nishimura, K., Agata, K., 2009. Planarians maintain a constant ratio of different cell types during changes in body size by using the stem cell system. Zool. Sci. 26, 805–813.

Takuma, N., et al., 1998. Formation of Rathke's pouch requires dual induction from the diencephalon. Development 125, 4835–4840.

Tanaka, S., 2007. Albino *corpus cardiacum* extracts induce morphometric gregarization in isolated albino locusts, *Locusta migratoria*, that are deficient in corazonin. Physiol. Entomol. 32, 95–98.

Thiel, D., Franz-Wachtel, M., Aguilera, F., Hejnol, A., 2018. Changes in the Neuropeptide Complement Correlate with Nervous System Architectures in Xenacoelomorphs bioRxiv 265579. Available from: https://www.biorxiv.org/content/biorxiv/early/2018/02/19/265579.full.pdf.

Thomas, Y., Bethenod, M.-T., Pelozuelo, L., Frerot, B., Bourguet, D., 2003. Genetic isolation between two sympatric host-plant races of the European corn borer, Ostrinia nubilalis Huebner. I. Sex pheromone, moth emergence timing, and parasitism. Evolution 57, 261–273.

Tomiczek, B.P., 2017. Computational Analysis of Gene Content in *Xenacoelomorpha*. University College, London. Thesis (Ph.D.).

Treier, M., O'Connell, S., Gleiberman, A., Price, J., Szeto, D.P., Burgess, R., et al., 2001. Hedgehog signaling is required for pituitary gland development. Development 128, 377–386.

Truman, J.W., 2006. Steroid hormone secretion in insects comes of age. Proc. Natl. Acad. Sci. U.S.A. 103, 8909–8910.

Tuisku, F., Hildebrand, C., 1994. Evidence for a neural influence on tooth germ generation in apolyphyodont species. Dev. Biol. 165, 1–9.

Uchida-Oka, N., Sugimoto, M., 2001. Norepinephrine induces apoptosis in skin melanophores by attenuating cAMP-PKA signals via alpha2-adrenoceptors in the medaka, *Oryzias latipes*. Pigment Cell Res 14, 356–361.

Umesono, Y., Agata, K., 2009. Evolution and regeneration of the planarian central nervous system. Dev. Growth Differ. 51, 185–195.

Ushatinskaya, G.T., 2008. Origin and dispersal of the earliest brachiopods. Paleontol. J. 42, 776–791.

Valentine, J.W., 1994. Late precambrian bilaterians: grades and clades. Proc. Natl. Acad. Sci. U.S.A. 91, 6751–6757.

Valentine, J.W., Jablonski, D., Erwin, D.H., 1999. Fossils, molecules and embryos: new perspectives on the Cambrian explosion. Development 126, 851–859.

Vannier, J., Calandra, I., Gaillard, C., Żylińska, A., 2010. Priapulid worms: pioneer horizontal burrowers at the Precambrian-Cambrian boundary. Geology 38, 711–714.

Vidal, B., Pasqualini, C., Le Belle, N.M., Holland, C.H., Sbaihi, M., et al., 2004. Dopamine inhibits luteinizing hormone synthesis and release in the juvenile European eel: a neuroendocrine lock for the onset of puberty. Biol. Reprod. 71, 1491–1500.

Wang, D.Y.-C., Kumar, S., Hedge, S.B., 1999. Divergence time estimates for the early history of animal phyla and the origin of plants, animals and fungi. Proc. R. Soc. Lond. B 266, 163–171.

Watanabe, H., Kuhn, A., Fushiki, M., Özbek, S., Fujisawa, T., et al., 2014. Sequential actions of â-catenin and Bmp pattern the oral nerve net in *Nematostella vectensis*. Nat. Commun. 5, 5536.

Weiss, L.C., Leimann, J., Tollrian, R., 2015. Predator-induced defences in *Daphnia longicephala*: location of kairomone receptors and timeline of sensitive phases to trait formation. J. Exp. Biol. 218, 2918–2926.

Wiens, J.J., Engstrom, T.N., Chippindale, P.T., 2006. Rapid diversification, incomplete isolation, and the "speciation clock' in North American Salamanders (Genus Plethodon): testing the hybrid swarm hypothesis of rapid radiation. Evolution 60, 2585–2603.

Williams, R.W., Herrup, K., 1988. Last revised 2001). The control of neuron number. Annu. Rev. Neurosci. 11, 423–453.

Wittbrodt, J., 2002. Brain patterning and eye development. EMBL Res. Rep. 93.

Wray, G.A., Levinton, J.S., Shapiro, L.H., 1996. Molecular evidence for deep precambrian divergences among metazoan phyla. Science 274, 568–573.

Wurtzel, O., Cote, L.E., Poirier, A., Satija, R., Regev, A., Reddien, P.W., 2015. A generic and cell-type-specific wound response precedes regeneration in planarians, 35, 632–645.

Xianguang, H., Bergstrom, J., 1997. Arthropods of the lower Cambrian Chengjiang fauna, southwest China. Foss. Strata 45, 1–116.

Yang, J., Ortega-Hernández, J., Butterfield, N.J., Liu, Y., Boyan, G.S., Hou, J-b., et al., 2016. Fuxianhuiid ventral nerve cord and early nervous system evolution in Panarthropoda. Proc. Natl. Acad. Sci. U.S.A. 113, 2988–2993.

Zhang, H., Hollander, J., Hansson, L.-A., 2017. Bi-directional plasticity: rotifer prey adjust spine length to different predator regimes. Sci. Rep. 7, 10254.

Zhang, Z.-Q., 2011. Animal biodiversity: an outline of higher-level classification and survey of taxonomic richness. Zootaxa 3147, 1–237.

Zhao, H., Feng, J., Seidel, K., Shi, S., Klein, O., Sharpe, P., et al., 2014. Secretion of Shh by a neurovascular bundle niche supports mesenchymal stem cell homeostasis in the adult mouse incisor. Cell Stem Cell 14, 160–173.

Zhuravlev, A.Y., Wood, R.A., 2018. The two phases of the Cambrian Explosion. Sci. Rep. 8, 16656.

Further reading

Bengtson, S., Budd, G., 2004. Comment on small bilaterian fossils from 40 to 55 million years before the cambrian. Science 306, 1291.

Bowring, S.A., Grotzinger, J.P., Condon, D.J., Ramezani, J., Newall, M., 2007. Geochronologic constraints on the chronostratigraphic framework of the Neoproterozoic Huqf Supergroup, Sultanate of Oman. Am. J. Sci. 307, 1097–1145.

Bowring, S.A., Grotzinger, J.P., Isachsen, C.E., Knoll, A.H., Pelechaty, S.M., Kolosov, P., 1993. Calibrating rates of early Cambrian evolution. Science 261, 1293–1298.

Brauchle, M., Bilican, A., Eyer, C., August 2018. Enacoelomorpha survey reveals that all 11 animal homeobox gene classes were present in the first bilaterians. Genome Biol. Evol. 10, 2205–2217.

Buatois, L.A., Mángano, M.G., 2016. Ediacaran ecosystems and the dawn of animals. In: Mángano, M., Buatois, L. (Eds.), The Trace-Fossil Record of Major Evolutionary Events, Topics in Geobiology, vol. 39. Springer, Dordrecht.

Cabej, N.R., 2012a. Epigenetic Principles of Evolution. Elsevier, London – Waltham, MA, pp. 147–204.

Cabej, N.R., 2013. Building the Most Complex Structure on Earth. Elsevier, London – Waltham, MA, pp. 121–192.

Cebrià, F., 2007. Regenerating the central nervous system: how easy for planarians! Dev. Genes Evol. 217, 733–748.

Fedonkin, M.A., Simonetta, A., Ivantsov, A.Y., 2007. New Data on *Kimberella*, the Vendian Mollusc-like Organism (White Sea Region, Russia): Palaeoecological and Evolutionary Implications, vol. 286. Geological Society, London, Special Publications, pp. 157–179.

Fedonkin, M.A., Waggoner, M., 1997. The Late Precambrian fossil *Kimberella* is a mollusc-like bilaterian organism. Science 388, 868–871.

Foote, M., 2003. Origination and extinction through the Phanerozoic: a new approach. J. Geol. 111, 125–148.

Hampe, W., Urny, J., Franke, I., Hoffmeister-Ullerich, S.A., Herrmann, D., Petersen, C.M., et al., 1999. A head-activator binding protein is present in hydra in a soluble and a membrane-anchored form. Development 126, 4077–4086.

Leger, S., Brand, M., 2002. Fgf8 and Fgf3 are required for zebrafish ear placode induction, maintenance and inner ear patterning. Mech. Dev. 119, 91–108.

Leitz, T., Morand, K., Mann, M., 1994. A novel peptide controlling development of the lower metazoan *Hydractinia echinata* (Coelenterata, Hydrozoa). Dev. Biol. 163, 440–446.

Lerner, U.H., 2002. Neuropeptidergic regulation of bone resorption and bone formation. J. Musculoskelet. Neuronal Interact. 2, 440–447.

Martin, M.W., Grazhdankin, D.V., Bowring, S.A., Evans, D.A.D., Fedonkin, M.A., Kirschvink, J.L., 2000. Age of neoproterozoic bilatarian body and trace fossils, white sea, Russia: implications for metazoan evolution. Science 288, 841–845.

Mitchell, N.C., Lin, J.I., Zaytseva, O., Cranna, N., Lee, A., Quinn, L.M., 2013. The ecdysone receptor constrains wingless expression to pattern cell cycle across the *Drosophila* wing margin in a cyclin B-dependent manner. BMC Dev. Biol. 13, 28.

Moroz, L.L., 2009. On the independent origins of complex brains and neurons. Brain Behav. Evol. 74, 177–190.

Nekrep, N., Wang, J., Miyatsuka, T., German, M.S., 2008. Signals from the neural crest regulate beta-cell mass in the pancreas. Development 135, 2151–2160.

Rosenkilde, P., Ussing, A.P., 1996. What mechanisms control neoteny and regulate induced metamorphosis in urodeles? Int. J. Dev. Biol. 40, 665–673.

Schaller, H.C., 1976. Action of the head activator as growth hormone in hydra. Cell Differ. 5, 1–11.

Schaller, H.C., 1976a. Action of the head activator on the determination of interstitial cells in hydra. Cell Differ. 5, 13–25.

Schaller, H.C., Bodenmuller, H., 1981. Isolation and amino acid sequence of a morphogenetic peptide from hydra. Dev. Biol. 78, 7000–7004.

Sepkoski, J.J., 1997. Biodiversity: past, present, and future. J. Paleontol. 71, 533–539.

Sepkoski Jr., J.J., September 16, 2002. A compendium of fossil marine animal genera. Bull. Am. Paleontol. 363, 1–560. http://taxonomicon.taxonomy.nl/Reference.aspx?id=5261.

Takeda, N., Konc, Y., Artigas, G.Q., Lapébi, P., Barreau, C., Koizumi, O., et al., 2018. Identification of jellyfish neuropeptides that act directly as oocyte maturation inducing hormones. Development 145, dev.156786.

Vargo, E.L., 1998. Primer pheromones in ants. In: Vander Meer, R.K., Breed, M.D., Espelie, K.E., Winston, M.L. (Eds.), Pheromone Communication in Social Insects. Ants, Wasps, Bees, and Termites. Westview Press, Boulder, CO, pp. 293–313.

Chapter 5

Epilogue

Chapter outline

Biological structures are patterned complex structures. The greater the complexity of a structure is, the higher the information requirements for erecting it are. Evolution of the immense number of species of ultracomplex structures during the Cambrian required the investment of unprecedented amounts of information.

The Cambrian explosion is a still unexplained, but not an unexplainable, process. The subject of this book was to infer the epigenetic mechanisms that triggered the Cambrian explosion by using my epigenetic hypothesis of metazoan evolution as an explanatory device.

It is extremely difficult to identify mechanisms that were operational half a billion years ago in living systems, of which we only can see and study their residuals, the trace and body fossils, which are not tractable at the physiological, behavioral, and only partly at the anatomical levels. Yet, it would be misleading to say that we are building on quicksand. To the contrary, there is a solid axiomatic theoretical ground: the Cambrian fauna is ancestral to the extant bilaterians.

Our general approach has been to go from the known present to the unknown past of metazoan life, by determining whether, and to what extent, we can use the relatively large volume of knowledge accumulated mainly during the last two centuries on the function and structure of extant animals to explain the mechanisms that set in motion the giant evolutionary machinery of the Cambrian explosion.

Epigenetic Mechanisms of the Cambrian Explosion. https://doi.org/10.1016/B978-0-12-814311-7.00005-6

Let us just succinctly mention in these closing remarks some of the important issues presented in the previous chapters and show how they can be applied to understanding and explaining the Cambrian explosion.

Epigenetic mechanisms of evolution and the source of the epigenetic information

Genes in the animal genome, according to their function, may be divided into the following:

- *Housekeeping genes*, whose products serve the maintenance of the normal composition and function of the cell, hence are constitutively expressed at a relatively constant rate in all the cells of the organism. These genes represent about 40% of the genes in human genome (Zhu et al., 2008) or about 8000 genes.
- *Tissue-specific* (nonhousekeeping) *genes*, which are expressed exclusively in particular cell types and tissues or in different stages of development. Their products (enzymes, hormones, secreted proteins, growth factors, neuromodulators, etc.) are secreted "on demand," from the cell into body fluids in response to extracellular signals, to maintain the normal functions, structures, and composition of the animal organism. Hence, their action occurs at supracellular or systemic levels. Products of tissue-specific genes are necessary for cell differentiation, histogenesis, organogenesis, and the maintenance of the metazoan structure against eroding consequences of the metabolic activity and performance of vital functions, as well as thermodynamic forces of erosion of the structure.

At the lower molecular level, the basic problems an animal faces during the evolution, development, cell differentiation, histogenesis, and organogenesis are related to the mechanism of determination of spatiotemporal patterns of gene expression and production of more protein types than the unmanipulated genome can provide.

The first topic, i.e., spatiotemporal (species-specific) regulation of gene expression, relies on at least four epigenetic processes:

- Spatiotemporal restriction of gene expression (Cabej, 2018d),
- Induction of epigenetic marks (DNA methylation and histone modification),
- Cooption of genes/gene regulatory networks (GRNs), and
- Regulation of miRNA expression patterns.

The second topic, the need for increasing the number and variability of proteins, involves two other epigenetic mechanisms:

- Gene splicing, and
- Somatic mutations.

Unquestionably, no bilaterian organism could develop and survive without having evolved at least part of the above epigenetic mechanisms.

Experimental evidence provided in Chapter 3 adequately demonstrates that all these epigenetic mechanisms (except for the somatic mutation, whose mechanism still remains unclear), in one form or another, are under neural control and are activated via neural signals, i.e., the information flows from the nervous system to the target organs/tissues/cells. Neural mechanisms of induction of the above-mentioned epigenetic changes represent the most parsimonious, indeed the only, mechanisms presented so far.

To a great extent, not only the structure and functions but also the developmental mechanisms of bilaterians, especially several model organisms of this group, are satisfactorily known in some details. While the early embryonic development is under the control and regulation of the parental (maternal and paternal) cytoplasmic factors, the developmental mechanisms of organogenesis unravel in the processes of the postphylotypic development, transgenerational inheritance, and regeneration.

The CNS has an instructive role in development and organogenesis of extant bilaterians

Cambrian explosion affected only bilaterians, which have since become the only animal group with complex organized structures, whose erection requires as a *sine qua non* an investment of specific information. In the process of biological organization, the cells of specific types and numbers have to be arranged in specific patterns to form particular structures at specific times in strictly determined sites of the body to form tissues, organs, and other structures. The probability that each cell could spontaneously and randomly find its place in the animal structure is zero. In metazoan development, this probability rises to 1, clearly indicating the investment/generation of the information of some sort for the billions to trillions of metazoan cells to find their location in the developing animal body.

This is an epigenetic information, obviously different from the information contained in the genes of the genome, which has no other known function than to store and relay information for protein biosynthesis. It induces cell differentiation, histogenesis, and organogenesis not only in the process of development and transgenerational inheritance but also in regeneration of lost body parts. The generator of this nongenetic or epigenetic information can be revealed by tracing back the cascade of chemical signals up to their ultimate source. This approach leads us nowhere else but to the central nervous system (CNS).

As pointed out earlier, the early embryonic development is under the epigenetic control of the parental (maternal and paternal) cytoplasmic factors. Organogenesis begins in the phylotypic stage with formation of the CNS. The three tables below succinctly, but adequately, illustrate the instructive role of the CNS in examples of organogenesis, transgenerational inheritance, and sympatric speciation:

a *The postphylotypic development* begins when the CNS has just become operational and sends chemical signals inducing the formation of somites, muscles (*Drosophila*), induction of head placodes, development of tetrapod limbs, formation of the optic cup and lens, otic cup, cochlea and inner ear, salivary (submandibular) gland, teeth induction and development, formation of the pituitary, architecture of the Langerhans islets in the pancreas, induction of the feather-inducing epidermis, airway branching and lung morphogenesis, arterial branching, morphological changes during puberty, body mass in insects and mammals, and so on (Table 5.1) (see also Section Instructive role of the nervous system in development, transgenerational inheritance, and speciation in Chapter 4).

TABLE 5.1 Role of the nervous system in bilaterian organogenesis.

No	Neurally determined organ development	Source
1	Induction of head placodes	Léger and Brand (2002); Chatterjee et al., 2010
2	Inner ear patterning	Leger and Brand, 2002
3	Development of the otic cup	Brigande et al. (2000)
4	Regulation of airway branching and morphogenesis in mammals and invertebrates	Bower et al. (2014)
5	Determination of the pancreatic islet architecture	Borden et al. (2013)
6	Formation of the pituitary gland	Takuma et al., 1998; Treier et al. (2001)
7	Tubulogenesis of the salivary gland	Nedvetsky et al. (2014)
8	Development of the submandibular (salivary) gland	Knox et al. (2010)
9	Patterning of the arterial branching	Mukoyama et al. (2002); Hogan et al. (2004); Mukoyama et al. (2005); Li et al. (2013)
10	Specification of adenohypophysis in zebrafish	Herzog et al. (2004)
11	Regulation of bone formation	Lundberg et al., 2001; Lerner (2002); Burt-Pichat et al. (2005)
12	Nerve patterning in *Xenopus laevis*	Herrera-Rincon et al. (2017)
13	Determination of sex differences in protandrous hermaphroditic fish	Elofsson et al. (1997)
14	Regulation of apoptosis (programmed cell death) in medaka fish *Oryzias latipes*	Sugimoto et al. (2000); Uchida-Oka and Sugimoto (2001)

TABLE 5.1 Role of the nervous system in bilaterian organogenesis.—cont'd

No	Neurally determined organ development	Source
15	Suggested role of pulmonary neuroendocrine cells in lung morphogenesis	Pan et al. (2006); Cutz et al. (2008)
16	Differentiation and proliferation of the developing and adult heart	Nebigil et al. (2000)
17	Nephrogenesis	Sariola et al. (1988); Sariola et al. (1989a); Sariola et al. (1989b)
18	Temporal regulation of the morphophysiological changes of puberty in vertebrates	Smith et al. (2006); Vidal et al. (2004); Riboni et al. (1998)
19	Teeth development	Tuisku and Hildebrand (1994); Hildebrand et al. (1995); Luukko et al. (1997); Fried et al. (2000); Løes et al., 2002; Kettunen et al. (2005); Zhao et al. (2014); Crucke et al. (2015)
20	Formation of dermomyotome and the feather-inducing dermis	Olivera-Martinez et al. (2001); Chang et al. (2004)
21	Activation of the lens gene regulatory network	Reza and Yasuda, 2004a; Reza and Yasuda, 2004b; Carl et al. (2002); Wittbrodt (2002)
22	Determination and maintenance of body mass in mammals	Adams et al. (2001); Levin and Dunn-Meynell (2000); Baeckberg et al., 2003
23	Induction of limb bud and limb development	Berggren et al. (2001); McCaffery and Dräger, 1994; Berggren et al., 1999
24	Formation of vestibule and cochlea in the inner ear	Riccomagno et al., 2005
25	Formation of the optic cup and lens	Kamachi et al., 1998

b *Transgenerational inheritance*: All the described cases of the transgenerational inheritance in metazoans are neurally determined. In this category belong predator-induced defenses (development of helmets and neckteeth) in rotifers, increased proportion of winged offspring in insects on detecting the cues of their predator or as a result of overcrowding, changes in the life histories of the offspring (production of male-only generation or sexual generation and diapause in rotifers in response to overcrowding), switching to the long life mode of the offspring in *Caenorhabditis elegans*, in response to scarce food/overcrowding, and induction of phase transition in the offspring of locusts.

A partial list of the described cases of neurally induced transgenerational inheritance is provided in Table 5.2 (see also the Section Instructive role of the nervous system in development, transgenerational inheritance, and speciation in Chapter 4).

Metamorphosis in mollusks, insects, and vertebrates is an even more clearly visible form of strict control of the postphylotypic development by the CNS.

c *Sympatric speciation:* According to the neo-Darwinian viewpoint, specia-tion requires geographical isolation of populations, at least until their reproductive and genetic isolation has occurred. Accordingly, practically the only form of speciation is allopatric speciation. The fact that the Cambrian diversification occurred in oceans where the geographical isolation is practically impossible represents a conundrum for the neo-Darwinian viewpoint. However, now we have an adequate number of cases of sympatric speciation, formation of species within the same area

TABLE 5.2 Role of the nervous system in inheritance of acquired characters.

	Neurally determined transgenerational inheritance of acquired characters without changes in genes	Source
1	Increased proportion of winged offspring in the pea aphid, *Acyrthosiphon pisum*	Dixon and Agarwala, 1999; Kunert and Weiser, 2003; Kunert et al., 2005
2	Increased proportion of winged offspring in cotton aphid, *Aphis gossypii* (Glover), for several sequential generations	Mondor et al., 2005
3	Formation of a helmet and the doubling of carapace's thickness in *Daphnia cucullata*	Agrawal et al., 1999; Laforsch and Tollrian, 2004
4	Production of male-only generation by mainly female populations of crustaceans of genus *Daphnia*	Rider et al., 2005; Eads et al., 2008; Miyakawa et al., 2014; Medlock Kakaley et al., 2017
5	Development of "neckteeth" by the crustacean *Daphnia pulex*	Agrawal et al., 1999; Barry, 2002; Miyakawa et al., 2010; Weiss et al., 2015
6	Induction of sexuality or diapause by females of several species of *Brachionus* rotifers to their offspring	Gilbert, 2004
7	Switching to long life mode of the nematode worm *Caenorhabditis elegans*	Lee and Ashrafi, 2008; Greer et al., 2011; Rechavi et al., 2014
8	Induction and inheritance to the offspring of phase transition in two locust species, *Schistocerca gregaria* and *Locusta migratoria*	Rogers et al., 2004; Claeys et al., 2006; Tanaka, 2007; Anstey et al., 2009; Badisco et al., 2011

from two different populations of an original parental species involving no geographical isolation. The mechanism of reproductive isolation of two populations in sympatry as a first step toward the formation of two species is a neurocognitive mechanism that drives particular individuals/population to mate only/preferably individuals that express particular phenotypes, although these individuals may be visually (morphologically) indistinguishable from the rest of individuals in the population. A partial list of cases of sympatric speciation is provided in Table 5.3.

Neurocognitive mechanisms of isolation and speciation observed in extant bilaterian taxa must have played an important role in speciation in the borderless oceans of the Cambrian era.

The maintenance of the metazoan structure

After the erection of the adult structure, the organism is faced with the equally difficult task of maintaining its structure against the thermodynamic and other forces of erosion and decomposition. Vital activities occurring in living organisms unavoidably produce metabolic noxious waste, leading to decline in the levels of the molecular components in body fluids and extracellular space and loss of cell components, which need to be detected in time and restored; otherwise the organism is faced with demise. To function normally, metazoan cells must live in, or have access to, the extracellular fluid from which they extract nutritive and other substances spent in the process of metabolism. They also have to replace millions of cells, of various types, continually lost.

From the beginning, metazoans, bilaterians included, evolved a system of control to monitor and detect changes in the molecular parameters of body fluids. In bilaterians, this function was taken over by the CNS. It is elementary physiological knowledge that the CNS controls not only the behavior but also the physiological functions of all the organs and organ systems (digestive, respiratory, excretory, circulatory, endocrine) and is responsible for maintaining the physiological levels of thousands of body fluid parameters. This implies the presence in the bilaterian organisms of a mechanism that continually monitors the level of these parameters in body fluids. When the values of these parameters, including body temperature (Hammel et al., 1963), drop below or exceed the physiological levels, the nerve endings send afferent information to the brain. The input is integrated and processed in respective brain areas generating the output in the form of a chemical signal that starts a signal cascade ultimately inducing the tissue-specific target cells to secrete the products that need to be replaced (proteins, hormones, enzymes, miRNAs, neuromodulators, etc.). Similarly, starting from planarians (Takeda et al., 2009), extremely simple bilaterians, insects and mammals, evolved a neural "counting mechanism" to maintain a relatively constant number of cells in adult life.

TABLE 5.3 Sympatric speciation based on neurocognitive mechanisms.

	Neurally determined reproductive isolation in sympatry and sympatric speciation	
1	Olfactory and auditory determined sympatric reproductive isolation in the Mexican pupfish	Strecker and Kodric-Brown, 1999
2	Olfactory and auditory determined sympatric reproductive isolation in *Drosophila*	Ortiz-Barrientos et al. (2004)
3	Acoustically determined sympatric speciation of *Gryllus texensis* and *Gryllus rubens*	Gray and Cade, 2000
4	Electrogenic maintenance of reproductive isolation of a group of five morphologically undistinguishable fish species of the genus *Campylomormyrus* in a Central African river basin	Hopkins, 1999
5	Use of electric wave forms for mate recognition by several morphologically indistinguishable morphs in sympatry	Arnegard et al., 2005; Arnegard et al., 2006
6	Neurocognitive speciation of the apple fruit fly, *Rhagoletis pomonella*	Linn et al., 2003; Linn et al., 2004
7	Two experimentally interbreeding races of the larch budmoth, *Zeiraphera diniana* (Lepidoptera: Tortricidae), do not interbreed in nature based on perception of their own race's mating signals and pheromones released by their females	Emelianov et al., 2003
8	Formation within last centuries of two sympatric species in the American continent from the European corn borer, *Ostrinia nubilalis* (Lepidoptera: Crambidae) by virtue of differences in male preferences for pheromones secreted by females and differences in the time of emergence between the two species	Bush, 1975; Thomas et al., 2003
9	A monophyletic group of forest dwelling Laupala crickets in Hawaii evolved with a speciation rate that was 26 times higher that the average speciation rate, based on evolution of female preference to mate conspecific males with courtship song's pulse rate characteristic for each species	Mendelson and Shaw, 2005
10	Mate choice based on the recognition of, and preference for, species-specific visual signals in formation of hundreds of species in Lake Victoria, Africa	Seehausen,1997a,b; Seehausen and van Alphen, 1998
11	Speciation via mate choice based on olfactory mate signals	Plenderleith et al., 2005
12	Speciation based on auditory mate signals	Amorim et al., 2004
13	Rapid speciation of 35 species of Plethodon salamanders in Northeastern United States during the last 5 million years and maintenance of reproductive isolation in sympatry based on the avoidance of interbreeding as a result of conspecific mate preferences	Wiens et al., 2006

This is an integrated control system, with the CNS as its controller, which continually compares the actual state with the "normal" physiological state, which is determined by the experimentally demonstrated existence in the CNS of "set points" (Fig. 5.1).

Animal organisms contain a specific number of cells which, under normal conditions, is maintained relatively constant throughout the life. There is empirical evidence that even a simple organism like a planarian has a neural "counting machine" that allows it to maintain a constant number of cells, but even when cut in pieces, each piece develops into a normal adult organism with the same number of cells (Takeda et al., 2009). The same is observed in insects. The growth in the freshwater snail *Lymnaea stagnalis* is controlled by a group of secretory neurons in the brain of the snail (Smit et al., 1998; Hatakeyama et al., 2000). Neural mechanisms also control the body mass in higher vertebrates and mammals (Pelleymounter et al., 1995; Halaas et al., 1995; Norman and Litwack, 1997, p. 240).

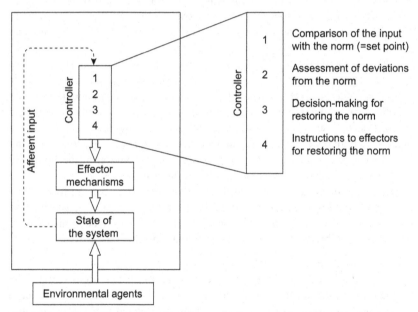

FIGURE 5.1 Generalized and simplified diagram of the integrated control system in metazoans with CNS acting as controller of system. A metazoan structure degrades continually due to metabolic activity, intrinsic thermodynamically determined causes, and a result of adverse influences of the environment. Changes in the structure and function of the organism and environmental changes are monitored by a pervasive network of interoceptors and exteroceptors and communicated to the controller. In the CNS, the input is compared with the neurally determined set points (1). Deviations from the norm are identified (2) and pathways for restoring the norm are determined (3). "Instructions" or commands for restoring the norm (4) are sent to effectors in target tissues and cells through signal cascades. Via the molecular and afferent feedback input, the controller receives continually information on the restored/degraded state of the system. *From Cabej, N.R., 2018c. Ibid. 55.*

What this and previous work proposed and substantiated with firmly established facts in this and previous works is that the nervous system plays an instructive role in the development and maintenance of animal morphology and organogenesis (Cabej, 2005; Cabej, 2012; Cabej, 2018).

The CNS and the Cambrian explosion

We have already seen that while the cnidarians and bilaterians share a basically common genetic toolkit, cnidarians became stuck in an evolutionary loop, while bilaterians produced the unprecedented Cambrian diversity of biological forms. The most conspicuous difference between the two groups is the centralization of the nervous system that occurred in bilaterians. This centralization and increase in the number of neurons enabled the bilaterian brain to generate the huge amount of information necessary for the development of organs and organ systems and the maintenance of the complex bilaterian structure. Obviously, this was impossible for the diffuse neural net situated between the ectoderm and endoderm throughout the cnidarian body (see also Section Centralization of the nervous system and the exponential increase of computational power in Chapter 4).

The role of the CNS in development (including metamorphosis), organogenesis, regeneration, and induction of the transgenerational inheritance is adequately demonstrated. Although the correlation and coincidence of the centralization of the nervous system with the triploblasty, advent of organs and organ systems, and the accelerated increase in structural complexity in bilateria is unlikely to be a historical accident, they do not warrant the drawing of inferences on a causal relationship between the evolution of the brain and the burst of bilaterian diversification during the Cambrian.

The event dealt with in this work occurred more than half a billion years ago, i.e., beyond real-time fact finding. However, a plethora of other scientific facts allow us to make reasonable inferences on the mechanism and drivers behind the Cambrian explosion. Our empirical and theoretical sources are:

- Facts on the developmental mechanisms of extant organisms,
- Trace and body fossils of the Cambrian era, which demonstrate the presence in the Cambrian of animals that, by consensus, are considered ancestral to extant taxa or represent linear descendants of Cambrian lineages, and
- General biological knowledge.

Now we have to examine whether, or to what extent, our knowledge on the development and evolution of extant animals can be extrapolated to the study and explanation of the Cambrian evolution.

In the following, a brief summary of firmly established facts supporting the hypothesis on the epigenetic mechanisms of the development of extant bilaterians and the possible inferences that may be drawn from the relevant empirical evidence are presented.

Justifying the extrapolation of the CNS as the source of information in development of extant taxa to the Cambrian diversification

What we can say with certainty is that there is no other way for any phenotypic change to become heritable and transmitted to the offspring, both in extant animals and in the Cambrian fauna, but via specific changes in the developmental pathways. Hence, the extrapolation of the knowledge of extant developmental pathways to the study of the evolution of the Cambrian fauna will depend to a great extent on the answer to the question: "have developmental pathways of extant animals been operational during the Cambrian?"

The core components of several basic developmental pathways have been conserved across the metazoan taxa since the Cambrian and beyond (Hinman et al., 2003). The fact that the main developmental pathways (Hedgehog [Hh], wingless-related [Wnt], transforming growth factor-β [TGF-β], receptor tyrosine kinase [RTK], Notch, Janus kinase [JAK]/signal transducer and activator of transcription [STAT], and nuclear hormone pathways) evolved even before the emergence of bilaterians and are conserved across the extant taxa (Pires-daSilva and Sommer., 2003) from planarians to vertebrates seems to indicate that they are conserved for evolutionarily long periods of time, that is, since the Cambrian evolution, when these taxa began to diverge from each other. Developmental pathways interact with each other becoming, thus, parts of greater regulatory networks (Bausek, 2013). A three-gene feedback loop for endoderm specification in sea urchin is conserved in the starfish now, more than half a billion years after their divergence during the Cambrian explosion (Hinman and Davidson, 2003).

In favor of the conservation of the ancestral and Cambrian developmental pathways witnesses also the fact that insects evolved wings even 300 million years after having lost them. Loss of wings in insects occurred thousands of times (Whiting et al., 2003), but the developmental pathways and GRNs for wing development was conserved even after the loss, as indicated by the repeated reevolution of insect wings.

The fossil record often shows striking examples of resemblance between Cambrian fossils (especially arthropods) and extant animals, a fact that "must imply retention of ancestral developmental GRN features" (Hinman et al., 2003).

The presence of so-called "living fossils," not only cnidarians and ctenophores, which have conserved their ancestral Bauplaene from times extending beyond the Cambrian, but also the arthropod-grade Cambrian *Kerygmachela kierkegaardi* gen et sp. nov. (Budd, 1993; Park et al., 2018) indicates that ancestral pathways in metazoans are conserved for long periods of time.

Horseshoe crabs are experiencing a long evolutionary stasis of at least 450 million years (Rudkin and Young, 2009), and the oldest coelacanth fish fossils found so far are dated 407–409 million years ago in Australia (Johanson et al., 2006), but its lineage, in all likelihood, originates in the

Cambrian era. The survival of these "living fossils" may be another proof that Cambrian developmental pathways are conserved.

Even when the phenotypic trait determined by a GRN does not develop, the developmental pathway is conserved and may be experimentally activated to produce the lost phenotype. The loss of teeth birds is a case in point. This class lost teeth about 60−80 million years ago (Chen et al., 2000), but experimentally teeth-like structures were induced in birds grafted with trans-specific (mouse) neural crest of prosencephalon origin (Kollar and Fisher, 1980).

Other examples of the conservation of ancestral developmental pathways after the loss of the structures they determine are the evolutionary reversion of shell coiling in gastropods (Collin and Cipriani, 2003), reversion of cartilaginous skeleton in sharks (Colbert and Morales, 1991), reversion of limbs in snakes (Tchernov et al., 2000), reversion of cartilaginous skeleton in fish (Colbert and Morales, 1991), experimental reversion of "hip glands" in moles *Microtus pennsylvanicus* and *Microtus longicaudus* (Jannett, 1975), and atavisms.

The fact that many of taxa belonging to the same lineages since the Cambrian have considerably diverged in various ways represents no real explanatory difficulty if we remember that, despite the general conservation, developmental pathways are flexible enough to be recruited to specify new tissues and organs (Pires-daSilva and Sommer, 2003) and recruit other genes. The conservation and flexibility of developmental pathways is paradigmatically exposed in cases of species that in the larval stage, i.e., with the same genetic toolkit and "the same developmental pathways," during their lifetime produce the Bauplan of another phylum/class (e.g., insect larvae build worm Bauplan and amphibian larvae develop fish organs such as whiskers, gills, lateral line sense organs, fins, etc.).

And as a final example of the conservation of the metazoan developmental pathways for long evolutionary and geological periods of time, one may mention the morphological similarity of pre-Cambrian era sponges and cnidarians with their extant forms.

Conservation of gene regulatory networks and the role of the CNS

In the last two decades, a shift occurred in the attention of most biologists from genes to GRNs as the regulators of gene expression, animal morphogenesis and development in general. Mapping GRNs reveals the causal relationships between the regulatory genes via detection of their spatiotemporal expression patterns, with the development of phenotypic traits, thus relating the genome with the development. Many of these GRNs are also conserved for long periods of time.

The conservation of the ancestral GRNs is illustrated with the similarity of GRNs that specify the endoderm in the sea urchin and the starfish, which diverged during the Cambrian, half a billion years ago (Hinman et al., 2003), but let's illustrate the conservation of GRNs with only two cases, the conservation of the eye GRN and insect wing GRN.

As many as ~2500 genes are involved in the process of the eye development, expression of *Pax6* and a number of other transcription factors (*Rx1*, *Lhx2*, *Six3*, and *Six6*) in a highly conserved GRN in retinal progenitor cells and are essential for the eye development to take place. The crucial role of *Pax6*, *Six2*, and *Ath*, and so on, is shared by almost all metazoans, including those possessing no eyes, beginning with Urbilateria (Arendt et al., 2002; Arendt and Baptista, 2002; Arendt et al., 2002).

This implies that the wide differences in the building plans of different eyes in metazoans result from the fact that different species, in the course of evolution, recruited different genes to the same ancestral eye developmental GRN. There is evidence suggesting that the CNS may be critically involved in gene recruitment (Cabej, 2011), and this has been hinted earlier by Nilsson:

> *When a simple nervous system became more elaborate, and new sensory organs were acquired, the ancient* Pax6 *genes gradually changed their role to include new targets which were functionally related to, or physically near, older targets.*
>
> Nilsson (1996)

Another example is formation of wings in *Drosophila melanogaster*. Wing development is a function of a GRN consisting of 468 genes (Li and White, 2003), i.e., 3.4% of 13,600 genes of the fly (Adams, 2000). In *Drosophila*, the insect wing GRN is conserved across holometabolous insects for about 300 million years (Abouheif and Wray, 2002).

The insect wings are suppressed by the expression of *Scr* (sex comb reduced) in the first thoracic segment but develop in the second segment (T2), where a gene that has no effect on wing development is expressed. In the third segment, T3, where the *Hox* gene *Ubx* (Ultrabithorax) is expressed, only halteres develop. It is demonstrated that the wing GRN is activated via secretion of ecdysone (Abouheif and Wray, 2002; Mitchell et al., 2013), which is neurally regulated, and the drop of the similarly neurally regulated JH (juvenile hormone) (Fig. 5.2).

Another example of conservation of the wing GRN in insects is that of the insects of the genus *Lopaphus*, which produce winged and unwinged offspring (Whiting et al., 2003).

The frequent reversion (Whiting et al., 2003) of wings in insects after their loss also illustrates conservation of the wing GRNs in the unwinged insects (Fig. 5.3).

While the conservation of the developmental pathways/GRNs in bilateria is a proven fact, the question rises about the mechanism that triggers their activation after long periods of inactivation. Are developmental pathways/GRNs self-regulated or their activation depends on factors that are external to them? As shown in this and previous work, signals for activation of developmental pathways/GRNs come from the CNS.

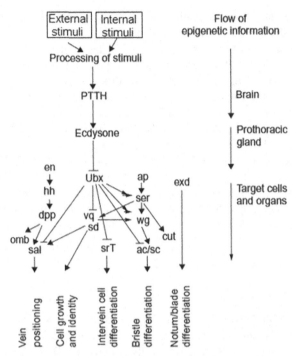

FIGURE 5.2 Flow of epigenetic information from the brain along ecdysone pathway for cell differentiation and growth in the process of wing development in insects. *PTTH,* prothoracicotropic hormone. *Based on Abouheif and Wray (2002). Slightly expanded from Cabej, N.R., 2018a. Epigenetic Principles of Evolution. Academic Press, London — San Diego Cambridge — Oxford, 401.*

It is amazing to think how the barely centralized nervous system of a planaria-like Cambrian animal, through a series of intermediate steps, rose to take over the control of not only the behavior and physiology but also the organogenesis and the maintenance of the metazoan structure. It is amazing to envisage how within half a billion years it evolved into the human brain, the most complex structure we know of. It is awesome to imagine how within the last 50—100,000 years the human brain changed the human being from an object of the living nature into a subject, from an unconscious to the only conscious being that rose from animal apathy to its environment to contemplation, intellectual curiosity, and ability to inquire about its own and its carrier's nature, the nature of life and our own existence! It is a wonderful and the most magnificent narrative to contepmplate how the evolution of the human brain pulled our own species out of the Animal Kingdom and transformed *Homo sapiens* into the real ruler of life on Earth.

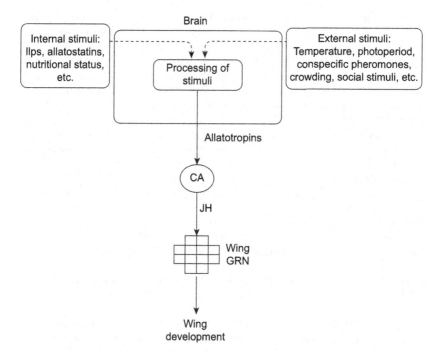

FIGURE 5.3 Generalized and simplified diagram of the wing development in response to internal and external stimuli in insects. *CA*, corpora allata; *GRN*, gene regulatory network; *JH*, juvenile hormone. *From Cabej, N.R., 2018b. Epigenetic Principles of Evolution. Academic Press, London — San Diego Cambridge — Oxford 408.*

References

Abouheif, E., Wray, G.A., 2002. Evolution of the gene network underlying wing polyphenism in ants. Science 297, 249–251.

Adams, M.D., 2000. The genome sequence of *Drosophila melanogaster*. Science 287, 2185–2195.

Adams, C.S., Korytko, A.I., Blank, J.L., 2001. A novel mechanism of body mass regulation. J. Exp. Biol. 204, 1729–1734.

Agrawal, A.A., Laforsch, C., Tollrian, R., 1999. Transgenerational induction of defences in animals and plants. Nature 401, 60–63.

Amorim, M.C.P., Knight, M.E., Stratoudakis, Y., Turner, G.F., 2004. Differences in sounds made by courting males of three closely related Lake Malawi cichlid species. J. Fish Biol. 65, 1358–1371.

Anstey, M.L., Rogers, S.M., Ott, S.R., Burrows, M., Simpson, S.J., 2009. Serotonin mediates behavioral gregarization underlying swarm formation in desert locusts. Science 323, 627–630.

Arendt, D., Baptista, I., 2002. Evolutionary aspects of eye and nervous system development. In: Wittbrodt, J. (Ed.), Brain Patterning and Eye Development. EMBL Research Reports, pp. 93–96, 2002.

228 Epigenetic Mechanisms of the Cambrian Explosion

Arendt, D., Tessmar, K., de Campos-Baptista, M.-I.M., Dorresteijn, A., Wittbrodt, J., 2002. Development of pigment-cup eyes in the polychaete *Platynereis dumerilii* and evolutionary conservation of larval eyes in Bilateria. Development 129, 1143−1154.

Arnegard, M.E., Bogdanowicz, S.M., Hopkins, C.D., 2005. Multiple cases of striking genetic similarity between alternative electric fish signal morphs in sympatry. Evolution 59, 324−343.

Arnegard, M.E., Jackson, B.S., Hopkins, C.D., 2006. Time-domain signal divergence and discrimination without receptor modification in sympatric morphs of electric fishes. J. Exp. Biol. 209, 2182−2198.

Badisco, L., Huybrechts, J., Simonet, G., Verlinden, H., Marchal, E., Huybrechts, R., et al., 2011. Transcriptome analysis of the desert locust central nervous system: production and annotation of a *Schistocerca gregaria* EST database. PLoS One 6 (3).

Baeckberg, M., Collin, M., Ovesjoe, M.L., Meister, B., 2003. Chemical coding of GABAB receptor immunoreactive neurones in hypothalamic regions regulating body weight. J. Neuroendocrinol. 15, 1−14.

Barry, M.J., 2002. Progress toward understanding the neurophysiological basis of predator-induced morphology in *Daphnia pulex*. Physiol. Biochem. Zool. 75, 179−186.

Bausek, N., 2013. JAK-STAT signaling in stem cells and their niches in *Drosophila*. JAK-STAT 2 (3), e25686.

Berggren, K., Ezerman, E.B., McCaffery, P., Forehand, C.J., 2001. Expression and regulation of the retinoicacid syntheticenzymeRALDH-2in the embryonic chicken wing. Dev. Dynam. 222, 1−16.

Berggren, K., McCaffery, P., Dräger, U., Forehand, C.J., 1999. Differential distribution of retinoic acid synthesis in the chicken embryo as determined by immunolocalization of the retinoic acid synthetic enzyme, RALDH-2. Dev. Biol. 210, 288−304.

Borden, P., Houtz, J., Leach, S.D., Kuruvilla, R., 2013. Sympathetic innervation during development is necessary for pancreatic islet architecture and functional maturation. Cell Rep. 4, 287−301.

Bower, D.V., Lee, H.-K., Lansford, R., Zinn, K., Warburton, D., Fraser, S.E., Jesudason, E.C., 2014. Airway branching has conserved needs for local parasympathetic innervation but not neurotransmission. BMC Biol. 12, 92.

Brigande, J.V., Kiernan, A.E., Gao, X., Iten, L.E., Fekete, D.M., 2000. Molecular genetics of pattern formation in the inner ear: do compartment boundaries play a role? Proc. Natl. Acad. Sci. U.S.A. 97, 11700−11706.

Budd, G., 1993. A Cambrian gilled lobopod from Greenland. Nature 364, 709−711.

Burt-Pichat, B., Lafage-Proust, M.H., Duboeuf, F., Laroche, N., Itzstein, C., Vico, L., et al., 2005. Dramatic decrease of innervation density in bone after ovariectomy. Endocrinology 146, 503−510.

Bush, G.L., 1975. Modes of animal speciation. Annu. Rev. Ecol. Syst. 6, 339−364.

Cabej, N.R., 2005. Neural Control of Development. Albanet, Dumont NJ.

Cabej, N.R., 2011. Neural control of gene recruitment in metazoans. Dev. Dyn. 240, 1−8.

Cabej, N.R., 2012. Epigenetic Principles of Evolution. Elsevier, London−Waltham.

Cabej, N.R., 2018a. Epigenetic Principles of Evolution. Academic Press, London − San Diego Cambridge − Oxford, p. 401.

Cabej, N.R., 2018b. Epigenetic Principles of Evolution. Academic Press, London − San Diego Cambridge − Oxford, p. 408.

Cabej, N.R., 2018c. Ibid., p. 55.

Cabej, N.R., 2018d. Ibid., pp. 230−235.

Carl, M., Loosli, F., Wittbrodt, J., 2002. Six3 inactivation reveals its essential role for the formation and patterning of the vertebrate eye. Development 129, 4057–4063.

Chang, C.-H., Jiang, T.X., Lin, C.M., Burrus, L.W., Chuong, C.M., Widelitz, R., 2004. Distinct Wnt members regulate the hierarchical morphogenesis of skin regions (spinal tract) and individual feathers. Mech. Dev. 121, 157–171.

Chatterjee, S., Kraus, P., Lufkin, T., 2010. A symphony of inner ear developmental control genes. BMC Genet. 11, 68.

Chen, Y., Zhang, Y., Jiang, T.-X., Barlow, A.J., St Amand, T.R., Hu, Y., et al., 2000. Conservation of early odontogenic signaling pathways in Aves. Proc. Natl. Acad. Sci. U.S.A. 97, 10044–10049.

Claeys, I., Breugelmans, B., Simonet, G., Van Soest, S., Sas, F., De Loof, A., Vanden Broeck, J., 2006. Neuroparsin transcripts as molecular markers in the process of desert locust (*Schistocerca gregaria*) phase transition. Biochem. Biophys. Res. Commun. 341, 599–606.

Colbert, E.H., Morales, M., 1991. Evolution of Vertebrates, vol. 44. Wiley-Liss.

Collin, R., Cipriani, R., 2003. Dollo's law and the re-evolution of shell coiling. Proc. R. Soc. B Biol. Sci. 270, 2551–2555.

Crucke, J., Van de Kelft, A., Huysseune, A., 2015. The innervation of the zebrafish pharyngeal jaws and teeth. J. Anat. 227, 62–71.

Cutz, E., Yeger, H., Pan, J., Ito, T., 2008. Pulmonary neuroendocrine cell system in health and disease. Curr. Respir. Med. Rev. 4, 174–186.

Dixon, A.F.G., Agarwala, B.K., 1999. Ladybird-induced life-history changes in aphids. Proc. R. Soc. Biol. Sci. 266, 1549–1553.

Eads, B.D., Andrews, J., Colbourne, J.K., 2008. Ecological genomics in *Daphnia*: stress responses and environmental sex determination. Heredity 100, 184–190.

Elofsson, U., Winberg, S., Francis, R.C., 1997. Sex differences in number of preoptic GnRH immunoreactive neurons in a protandrously hermaphroditic fish, the anemone fish *Amphiprion melanopus*. J. Comp. Physiol. A 181, 484–492.

Emelianov, I., Simpson, F., Narang, P., Mallet, J., 2003. Host choice promotes reproductive isolation between host races of the larch budmoth *Zeiraphera diniana*. J. Evol. Biol. 16, 208–218.

Fedonkin, M.A., Simonetta, A., Ivantsov, A.Y., 2007. In: New Data on Kimberella, the Vendian Mollusc-Like Organism (White Sea region, Russia): Palaeoecological and Evolutionary Implications, 286. Geological Society, London, Special Publications, pp. 157–179.

Fried, K., Nosrat, C., Lillesaar, C., Hildebrand, C., 2000. Molecular signaling and pulpal nerve development. Crit. Rev. Oral Biol. Med. 11, 318–332 (Abstract).

Gilbert, J.J., 2004. Population density, sexual reproduction and diapause in monogonont rotifers: new data for *Brachionus* and a review. J. Limnol. 63 (Suppl. 1), 32–36.

Gray, D.A., Cade, W.H., 2000. Sexual selection and speciation in field crickets. Proc. Natl. Acad. Sci. U.S.A. 97, 14449–14454.

Greer, E.L., Maures, T.J., Ucar, D., Hauswirth, A.G., Mancini, E., Lim, J.P., Benayoun, B.A., Shi, Y., Brunet, A., 2011. Transgenerational epigenetic inheritance of longevity in *Caenorhabditis elegans*. Nature 479, 365–371.

Halaas, J.L., Gajiwala, K.S., Maffei, M., Cohen, S.L., Chait, B.T., Rabinowitz, D., et al., 1995. Weight-reducing effects of the plasma protein encoded by the obese gene. Science 269, 543–546.

Hammel, H.T., Jackson, D.C., Stolwijk, J.A.J., Hardy, J.D., Stroeme, S.B., 1963. Temperature regulation by hypothalamic proportional control with an adjustable set point. J. Appl. Phys. 18, 1146–1154.

Hatakeyama, D., Ito, I., Kojima, S., Fujito, Y., Ito, E., 2000. Complement receptor 3-like immunoreactivity in the light green cells and the canopy cells of the pond snail, *Lymnaea stagnalis*. Brain Res. 865, 102−106.

Herrera-Rincon, C., Pai, V.P., Moran, K.M., Lemire, J.M., Levin, M., 2017. The brain is required for normal muscle and nerve patterning during early Xenopus development. Nat. Commun. 8, 587.

Herzog, W., Sonntag, C., von der Hardt, S., Roehl, H.H., Varga, Z.M., Hammerschmidt, M., 2004. Fgf3 signalling from the ventral diencephalon is required for early specification and subsequent survival of the zebrafish adenohypophysis. Development 131, 3681−3692.

Hildebrand, C., Fried, K., Tuisku, F., Johansson, C.S., 1995. Teeth and tooth nerves. Prog. Neurobiol. 45, 165−222.

Hinman, V.F., Davidson, E.H., 2003. Expression of AmKrox, a starfish ortholog of a sea urchin transcription factor essential for endomesodermal specification. Gene Expr. Patterns 3, 423−426.

Hinman, V.F., Nguyen, A.T., Cameron, R.A., Davidson, E.H., 2003. Developmental gene regulatory network architecture across 500 million years of echinoderm evolution. Proc. Natl. Acad. Sci. U.S.A. 100, 13356−13361.

Hogan, K.A., Ambler, C.A., Chapman, D.L., Bautch, V.L., 2004. The neural tube patterns vessels developmentally using the VEGF signaling pathway. Development 131, 1503−1513.

Hopkins, C.D., 1999. Signal evolution in electric communication. In: Hauser, M.D., Konishi, M. (Eds.), The Design of Animal Communication. MIT Press, Cambridge, MA, pp. 461−491.

Jannett Jr., F.J., 1975. "Hip glands" of *Microtus pennsylvanicus* and M. *longicaudus* (Rodentia: muridae), voles "without" hip glands. Syst. Zool. 24, 171−175.

Johanson, Z., Long, J.A., Talent, J.A., Janvier, P., Warren, J.W., 2006. Oldest coelacanth, from the early Devonian of Australia. Biol. Lett. 2 (3), 443−446.

Kamachi, Y., Uchikawa, M., Collignon, J., Lovell-Badge, R., Kondoh, H., 1998. Involvement of Sox1, 2 and 3 in the early and subsequent molecular events of lens induction. Development 125, 2521−2532.

Kettunen, P., Løes, S., Furmanek, T., Fjeld, K., Kvinnsland, I.H., Behar, O., Yagi, T., et al., 2005. Coordination of trigeminal axon navigation and patterning with tooth organ formation: epithelial-mesenchymal interactions, and epithelial Wnt4 and Tgfβ1 regulate semaphorin 3a expression in the dental mesenchyme. Development 132, 323−334.

Knox, S.M., Lombaert, I.M.A., Reed, X., Vitale-Cross, L., Gutkind, J.S., Hoffman, M.P., 2010. Parasympathetic innervation maintains epithelial progenitor cells during salivary organogenesis. Science 329, 1645−1647.

Kollar, E.J., Fisher, C., 1980. Tooth inductionin chick epithelium: expression of quiescent genes for enamel synthesis. Science 207, 993−995.

Kunert, G., Weisser, W.W., 2003. The interplay between density- and trait-mediated effects in predatory-prey interactions: a case study in aphid wing polymorphism. Oecologia 135, 304−312.

Kunert, G., Weisser, W.W., 2005. The importance of antennae for pea aphid wing induction in the presence of natural enemies. B. Entomol. Res. 95, 125−131.

Kunert, G., Otto, S., Roese, U.S.R., Gershenzon, J., Weisser, W.W., 2005. Alarm pheromone mediates production of winged dispersal morphs in aphids. Ecol. Lett. 8, 596−603.

Laforsch, C., Tollrian, R., 2004. Extreme helmet formation in *Daphnia cucullata* induced by small scale turbulence. J. Plankton Res. 26, 81−87.

Lee, B.H., Ashrafi, K., 2008. A TRPV channel modulates, *C. elegans* neurosecretion, larval starvation survival, and adult lifespan. PLoS Genet. 4 https://doi.org/10.1371/journal.pgen.1000213.

Léger, S., Brand, M., 2002. Fgf8 and Fgf3 are required for zebrafish ear placode induction, maintenance and inner ear patterning. Mech. Dev. 119, 91–108.

Lerner, U.H., 2002. Neuropeptidergic regulation of bone resorption and bone formation. J. Musculoskelet. Neuronal Interact. 2, 440–447.

Levin, B.E., Dunn-Meynell, A.A., 2000. Sibutramine alters the central mechanisms regulating the defended body weight in diet-induced obese rats. Am. J. Physiol. Reg. I. 279, R2222–R2228.

Li, W., Kohara, H., Uchida, Y., James, J.M., Soneji, K., Cronshaw, D.G., et al., 2013. Peripheral nerve-derived CXCL12 and VEGF-A regulate the patterning of arterial vessel branching in developing limb skin. Dev. Cell 24, 359–371.

Li, T.-R., White, K.P., 2003. Tissue-specific gene expression and ecdysone-regulated genomic networks in *Drosophila*. Dev. Cell 5, 59–72.

Linn, C., Feder, J.L., Nojima, S., Dambroski, H.R., Berlocher, S.H., Roelofs, W., 2003. Fruit odor discrimination and sympatric host race formation in Rhagoletis. Proc. Natl. Acad. Sci. U.S.A. 100, 11490–11493.

Linn, C.E., Dambroski, H.R., Feder, J.L., Berlocher, S.H., Nojima, S., Roelofs, W.L., 2004. Postzygotic isolating factor in sympatric speciation in *Rhagoletis* flies: reduced response of hybrids to parental host-fruit odors. Proc. Natl. Acad. Sci. U.S.A. 101, 17753–17758.

Løes, S., Kettunen, P., Kvinnsland, H., Luukko, K., 2002. Mouse rudimentary diastema tooth primordia are devoid of peripheral nerve fibers. Anat. Embryol. (Berl.). 205, 187–191.

Lundberg, P., Lundgren, I., Mukohyama, H., Lehenkari, P.P., Horton, M.A., Lerner, U.H., 2001. Vasoactive intestinal peptide (VIP)/pituitary adenylate cyclase-activating peptide receptor subtypes in mouse calvarial osteoblasts: presence of VIP-2 receptors and differentiation-induced expression of VIP-1 receptors. Endocrinology 142, 339–347.

Luukko, K., Sariola, H., Saarma, M., Thesleff, I., 1997. Localization of nerve cells in the developing rat tooth. J. Dent. Res. 76, 1350–1356.

Martin, M.W., Grazhdankin, D.V., Bowring, S.A., Evans, D.A.D., Fedonkin, M.A., Kirschvink, J.L., 2000. Age of Neoproterozoic Bilatarian body and trace fossils, white sea, Russia: implications for metazoan evolution. Science 288, 841–845.

McCaffery, P., Dräger, U.C., 1994. Hot spots of retinoic acid synthesis in the developing spinal cord. Proc. Natl. Acad. Sci. U.S.A. 91, 7194–7197.

Medlock Kakaley, E.K., Wang, H.Y., LeBlanc, G.A., 2017. Agonist-mediated assembly of the crustacean methyl farnesoate receptor. Sci. Rep. 7.

Mendelson, T.C., Shaw, K.L., 2005. Rapid speciation in an arthropod. Nature 433, 375–376.

Mitchell, N.C., Lin, J.I., Zaytseva, O., Cranna, N., Lee, A., Quinn, L.M., 2013. The ecdysone receptor constrains wingless expression to pattern cell cycle across the *Drosophila* wing margin in a cyclin B-dependent manner. BMC Dev. Biol. 13, 28.

Miyakawa, H., Imai, M., Sugimoto, N., Ishikawa, Y., Ishikawa, A., Ishigaki, H., et al., 2010. Gene up-regulation in response to predator kairomones in the water flea, *Daphnia pulex*. BMC Dev. Biol. 10, 45.

Miyakawa, H., Toyota, K., Sumiya, E., Iguchi, T., 2014. Comparison of JH signaling in insects and crustaceans. Curr. Opin. Insect Sci. 1, 81–87.

Mondor, E.B., Rosenheim, J.A., Addicott, J.F., 2005. Predator-induced transgenerational phenotypic plasticity in the cotton aphid. Oecologia 142, 104–108.

Mukoyama, Y., Gerber, H.-P., Ferrara, N., Gu, C., Anderson, D.J., 2005. Peripheral nerve-derived VEGF promotes arterial differentiation via neuropilin 1-mediated positive feedback. Development 132, 941–952.

Mukoyama, Y.S., Shin, D., Britsch, S., Taniguchi, M., Anderson, D.J., 2002. Sensory nerves determine the pattern of arterial differentiation and blood vessel branching in the skin. Cell 109, 693–705.

Nebigil, C.G., Choi, D.-S., Dierich, A., Hickel, P., Le Meur, M., Messaddeq, N., et al., 2000. Serotonin 32B receptor is required for heart development. Proc. Natl. Acad. Sci. U.S.A. 97, 9508–9513.

Nedvetsky, P.I., Emmerson, E., Finley, J.K., Ettinger, A., Cruz-Pacheco, N., Prochazka, J., et al., 2014. Parasympathetic innervation regulates tubulogenesis in the developing salivary gland. Dev. Cell 30, 449–462.

Nilsson, D.-E., 1996. Eye ancestry: old genes for new eyes. Curr. Biol. 6, 39–42.

Norman, A.W., Litwack, G., 1997. Hormones, second ed. Academic Press, Boston, MA, p. 240.

Olivera-Martinez, I., Thélu, J., Teillet, M.-A., Dhouailly, D., 2001. Dorsal dermis development depends on a signal from the dorsal neural tube, which can be substituted by Wnt-1. Mech. Dev. 100, 233–244.

Ortiz-Barrientos, D., Counterman, B.A., Noor, M.A., 2004. The genetics of speciation by reinforcement. PLoS Biol. 2, e416.

Pan, J., Copland, I., Post, M., Yeger, H., Cutz, E., 2006. Mechanical stretch-induced serotonin release from pulmonary neuroendocrine cells: implications for lung development. Am. J. Physiol. Lung Mol. Physiol. 290, L185–L193.

Park, T.-Y.S., Kihm, J.-H., Woo, J., Park, C., Lee, W.Y., Smith, M.P., Harper, D.A.T., Young, F., Nielsen, A.T., 2018. Brain and eyes of *Kerygmachela* reveal protocerebral ancestry of the panarthropod head. Nat. Commun. 9 (1).

Pelleymounter, M.A., Cullen, M.J., Baker, M.B., Hecht, R., Winters, D., Boone, T., Collins, F., 1995. Effects of the obese gene product on body weight regulation in ob/ob mice. Science 269, 540–543.

Pires-daSilva, A., Sommer, R.J., 2003. The evolution of signaling pathways in animal development. Nat. Rev. Genet. 4, 39–49.

Plenderleith, M., van Oosterhout, C., Robinson, R.L., Turner, G.F., 2005. Female preference for conspecific males based on olfactory cues in a Lake Malawi cichlid fish. Biol. Lett. 1, 411–414.

Rechavi, O., Houri-Ze'evi, L., Anava, S., Goh, W.S.S., Kerk, S.Y., Hannon, G.J., Hobert, O., 2014. Starvation-induced transgenerational inheritance of small RNAs in *C. elegans*. Cell 158, 277–287.

Reza, H.M., Yasuda, K., 2004a. Lens differentiation and crystallin regulation: a chick model. Int. J. Dev. Biol. 48, 805–817.

Reza, H.M., Yasuda, K., 2004b. The involvement of neural retina Pax6 in lens fiber differentiation. Dev. Neurosci. 26, 318–327.

Riboni, L., Escamilla, C., Chavira, R., Dominguez, R., 1998. Effects of peripheral sympathetic denervation induced by guanethidine administration on the mechanisms regulating puberty in the female Guinea pig. J. Endocrinol. 156, 91–98.

Riccomagno, M.M., Takada, S., Epstein, D.J., 2005. Wnt-dependent regulation of inner ear morphogenesis is balanced by the opposing and supporting roles of Shh. Genes Dev. 19, 1612–1623.

Rider, C.V., Gorr, T.A., Olmstead, A.W., Wasilak, B.A., LeBlanc, G.A., 2005. Stress signaling: coregulation of hemoglobin and male sex determination through a terpenoid signaling pathway in a crustacean. J. Exp. Biol. 208, 15–23.

Rogers, S.M., Matheson, T., Sasaki, K., Kendrick, K., Simpson, S.J., Burrows, M., 2004. Substantial changes in central nervous system neurotransmitters and neuromodulators accompany phase change in the locust. J. Exp. Biol. 207, 3603–3617.

Rudkin, D.M., Young, G.M., 2009. Horseshoe crabs—an ancient ancestry revealed. In: Tanacredi, J.T., Botton, M.L., Smith, D.R. (Eds.), Biology and Conservation of Horseshoe Crabs. Springer, New York, pp. 25–44.

Sariola, H., Ekblom, P., Henke-Fahle, S., 1989a. Embryonic neurons as in vitro inducers of differentiation of nephrogenic mesenchyme. Dev. Biol. 132, 271–281.

Sariola, H., Holm, K., Heinke-Fahle, S., 1988. Early innervation of the metanephric kidney. Development 104, 589–599.

Sariola, H., Holm-Sainio, K., Henke-Fahle, S., 1989b. The effect of neuronal cells on kidney differentiation. Int. J. Dev. Biol. 33, 149–155.

Seehausen, O., van Alphen, J.J.M., 1998. The effect of male coloration on female mate choice in closely related Lake Victoria cichlids (*Haplochromis nyererei* complex). Behav. Ecol. Sociobiol. 42, 1–8.

Seehausen, O., van Alphen, J.J.M., Witte, F., 1997a. Cichlid fish diversity threatened by eutrophication that curbs sexual selection. Science 277, 1808–1811.

Seehausen, O., Witte, F., Katunzi, E.F.B., Smits, J., Bouton, N., 1997b. Patterns of the remnant cichlid fauna in southern Lake Victoria. Conserv. Biol. 11, 890–905.

Smit, A.B., van Kesteren, R.E., Li, K.W., van Heerikhuizen, H., van Minnen, J., Spijker, S., Geraerts, W.P.M., 1998. Towards understanding the role of insulin in the brain: lessons from insulinrelated signaling systems in the invertebrate brain. Prog. Neurobiol. 54, 35–54.

Smith, J.T., Clifton, D.K., Steiner, R.A., 2006. Regulation of the neuroendocrine reproductive axis by kisspeptin-GPR54 signaling. Reproduction 131, 623–630.

Strecker, U., Kodric-Brown, A., 1999. Mate recognition systems in a species flock of Mexican pupfish. J. Evol. Biol. 12, 927–935.

Sugimoto, M., Uchida, N., Hatayama, M., 2000. Apoptosis in skin pigmentcells of medaka, *Oryzias latipes* (Teleostei), during long-term chromatic adaptation: the role of sympathetic innervation. Cell Tissue Res. 301, 205–216 (Abstract).

Takeda, H., Nishimura, K., Agata, K., 2009. Planarians maintain a constant ratio of different cell types during changes in body size by using the stem cell system. Zool. Sci. 26, 805–813.

Takuma, N., et al., 1998. Formation of Rathke's pouch requires dual induction from the diencephalon. Development 125, 4835–4840.

Tanaka, S., 2007. Albino corpus cardiacum extracts induce morphometric gregarization in isolated albino locusts, *Locusta migratoria*, that are deficient in corazonin. Physiol. Entomol. 32, 95–98.

Tchernov, E., Rieppel, O., Zaher, H., Polcyn, M.J., Jacobs, L.L., 2000. A fossil snake with limbs. Science 287, 2010–2012.

Thomas, Y., Bethenod, M.-T., Pelozuelo, L., Frérot, B., Bourguet, D., 2003. Genetic isolation between two sympatric host-plant races of the European corn borer, *Ostrinia nubilalis* Huebner. I. Sex pheromone, moth emergence timing, and parasitism. Evolution 57, 261–273.

Treier, M., O'Connell, S., Gleiberman, A., Price, J., Szeto, D.P., Burgess, R., et al., 2001. Hedgehog signaling is required for pituitary gland development. Development 128, 377–386.

Tuisku, F., Hildebrand, C., 1994. Evidence for a neural influence on tooth germ generation in apolyphyodont species. Dev. Biol. 165, 1—9.

Uchida-Oka, N., Sugimoto, M., 2001. Norepinephrine induces apoptosis in skin melanophores by attenuating cAMP-PKA signals via alpha2-adrenoceptors in the medaka, *Oryzias latipes*. Pigm. Cell Res. 14, 356—361.

Vidal, B., Pasqualini, C., Le Belle, N.M., Holland, C.H., Sbaihi, M., et al., 2004. Dopamine inhibits luteinizing hormone synthesis and release in the juvenile European eel: a neuroendocrine lock for the onset of puberty. Biol. Reprod. 71, 1491—1500.

Weiss, L.C., Leimann, J., Tollrian, R., 2015. Predator-induced defences in *Daphnia longicephala*: location of kairomone receptors and timeline of sensitive phases to trait formation. J. Exp. Biol. 218, 2918—2926.

Whiting, M.F., Bradler, S., Maxwell, T., 2003. Loss and recovery of wings in stick insects. Nature 421, 264—267.

Wiens, J.J., Brandley, M.C., Reeder, T.W., 2006. Why does a trait evolve multiple times within a clade? Repeated evolution of snakelike body form in squamate reptiles. Evolution 60, 123—141.

Wittbrodt, J., 2002. Brain patterning and eye development. EMBL Res. Rep. 93.

Zhao, H., Feng, J., Seidel, K., Shi, S., Klein, O., Sharpe, P., et al., 2014. Secretion of Shh by a neurovascular bundle niche supports mesenchymal stem cell homeostasis in the adult mouse incisor. Cell Stem Cell 14, 160—173.

Zhu, J., He, F., Song, S., Wang, J., Yu, J., 2008. How many human genes can be defined as housekeeping with current expression data? BMC Genomics 9, 172.

Further reading

Consoulas, C., Levine, R.B., 1997. Accumulation and proliferation of adult leg muscle precursors in *Manduca* are dependent on innervation. J. Neurobiol. 32, 531—553.

Currie, D.A., Bate, M., 1995. Innervation is essential for the development and differentiation of a sex-specific adult muscle in *Drosophila melanogaster*. Development 121, 2549—2557.

Fernandes, J.J., Keshishian, H., 1998. Nerve-muscle interactions during flight muscle development in *Drosophila*. Development 125, 1769—1779.

Lawrence, P.A., Johnston, P., 1986. The muscle pattern of a segment of *Drosophila* may be determined by neurons and not by contributing myoblasts. Cell 45, 505—513.

Mitsiadis, T.A., Chéraud, Y., Sharpe, P., Fontaine-Pérus, J., 2003. Development of teeth in chick embryos after mouse neural crest transplantations. Proc. Natl. Acad. Sci. U.S.A. 100, 6541—6545.

Index

Printed in the United States
By Bookmasters